The Nautical Institute

Guidelines for Collecting Maritime Evidence

Guidelines for Collecting Maritime Evidence

Published by The Nautical Institute
202 Lambeth Road, London SE1 7LQ, England
Tel: +44 (0)20 7928 1351   Fax: +44 (0)20 7401 2817   Web: www.nautinst.org
© The Nautical Institute 2017

All rights reserved. No part of this publication may be reproduced, stored in a retrieval system, or transmitted in any form or by any means, electronic, mechanical, photocopying, recording or otherwise, without the prior written consent of the publisher, except for quotation of brief passages in reviews.

Although great care has been taken with the writing of the book and the production of the volume, neither The Nautical Institute nor the contributors can accept any responsibility for errors and omissions or their consequences.

This book has been prepared to address the subject of guidelines for collecting maritime evidence. This should not, however, be taken to mean that this document deals comprehensively with all of the concerns that will need to be addressed or even, where a particular matter is addressed, that this document sets out the only definitive view for all situations. The opinions expressed are those of the contributors only and are not necessarily to be taken as the policies or views of any organisation with which they have any connection.

Readers of *Guidelines for Collecting Maritime Evidence* are advised to make themselves aware of any applicable local, national or international legislation or administrative requirements or advice which may affect decisions taken on board.

Cover image ITOPF
Publisher Bridget Hogan
Book Editor Margaret Freeth
Typesetting and layout by Phil McAllister Design
Printed in the UK by Cambrian Printers, Aberystwyth
ISBN 978 1 906915 54 4

Acknowledgements

This book goes beyond the ship and those on board (covered in the accompanying volume *The Mariner's Role in Collecting Evidence – Handbook*) to look at the collection of evidence by the wide range of interests that may be involved after an accident or incident. The Nautical Institute would like to thank all those who agreed to share their knowledge and expertise, either as authors, peer reviewers or by giving advice and support. Their time and effort has produced sound and practical guidance.

Special thanks are due to the following:

The Technical Editor, John Noble BSc FNI, who consulted widely, decided the contents of the book, recruited the authors, technically reviewed their contributions and managed the peer review process.

The Institute's Technical Committee for advice on the issues most affecting members.

Bridget Hogan, Director of Publishing and Marketing, and Margaret Freeth, Book Editor, for assistance to the Technical Editor.

Jennefer Tobin MBE for information on developments in the collection and use of digital data.

Captain Zarir Irani FICS FIIMS AFNI NAMS-CMS HCMM AVI MBA for his support for the book.

Our peer reviewers: Nick Ferguson-Gow ACII; Richard Gunn Partner/Master Mariner; David Handley; Ben Harris LLB; Bruce Harris Chartered Arbitrator FCIArb; Graham Harris; Oessur Jarleivson Hilduberg, Head of DMAIB; Captain Zarir Irani FICS FIIMS AFNI NAMS-CMS HCMM AVI MBA; Captain C R Kelso MBE FNI; Jeremy Russell QC; Garry Stevens BSc Master Mariner; Captain Paul Voisin FNI AFRIN.

Captain Allen Brink FNI FRIN FIIMS FCMS HCMM and Karen Purnell PhD FRSC, Managing Director of ITOPF, for supplying images.

Authors would also like to thank their peer reviewers and the following:

Expedo Ship Management; V.Ships Compliance Department; V.Scope Risk Management Ltd; LMAA arbitrators Brian Williamson BSc CArb FRMetS FNI FEI FCIArb and Michael Allen BSc CEng MIMarEST; LMAA Honorary Secretary, Daniella Horton BSc MCIArb.

The publications team at The Nautical Institute would also like to acknowledge the pioneering work of the Institute's North East England branch and Dr Phil Anderson DProf FNI in producing earlier editions.

Foreword

By Philip Wake OBE MSc FNI
Chief Executive, The Nautical Institute

There are many reasons why maritime evidence needs to be collected and many different recipients for that evidence. The process of compiling evidence begins with the regular record keeping on board ship and in the management office ashore, whether in compliance with the ISM Code or the many regulations and procedures in place. If this is being done well, the task of putting together the evidence required after an accident or incident is made considerably easier. Similarly, the increasing availability of electronic data capture and transmitting systems on board ship are making it easier to compile and analyse data. It is to be hoped that companies will make far greater use of these systems in order to reduce the administrative burden on their sea staff and to provide the necessary proof of compliance.

However, what evidence needs to be collected, why and for whom? This timely book, and its companion volume, *The Mariner's Role in Collecting Evidence – Handbook*, set out to answer these questions in a concise but expert fashion. Written by experienced professionals in their fields and covering the range of interested parties involved in accident investigation, it is made accessible to all by the use of maritime English. It avoids lengthy chapters while still providing the essential information.

It is crucial that both volumes are made available to company fleets and shore staff so that all are fully aware of what will happen and what they will be required to produce as evidence should an accident or incident occur. Naturally, we all hope that we are not faced with a casualty situation but it is doubtful if there is a shipping professional anywhere who has gone through their entire service without being involved in some form of accident. When something does go wrong, it is not the time – and there is no time – to start reading about what you should collect in the way of evidence. That task must be undertaken when work is proceeding smoothly and as part of your normal professional development. This book should be required reading for all officers and not just held in the company's office for reference.

The authors of the chapters are to be commended for a job well done and John Noble has expertly pulled it all together as Technical Editor, using his immense experience in the maritime industry, including his leadership of the Salvage Association. It is to be hoped that this work will help to protect mariners from criminalisation after accidents because they will have recorded their actions accurately and thereby provided themselves with a good defence.

Dedication

This book is dedicated to the many seafarer organisations that step in to support mariners following an incident, often while the collection of evidence is under way.

Contents

Chapter 1: Introduction 1
By John Noble

Chapter 2: Preserving evidence on behalf of state safety inspectors: the UK perspective ... 5
By Captain Andrew Moll

Chapter 3: The Master's responsibilities for collecting evidence 13
By Captain Ian Odd

Chapter 4: The P&I approach 23
By Chris Adams

Chapter 5: The surveyor's perspective 29
By Captain Sanjay Bhasin

Chapter 6: The lawyer's point of view 39
By Michael Mallin and Jack Hatcher

Chapter 7: A marine claims broker's perspective 47
By David Keyes and Ivor Goveas

Chapter 8: Evidence for insurance claims 53
By Peter Young

Chapter 9: A practical view from P&I 65
By Louise Hall

Chapter 10: Evidence and arbitration 71
Captain Julian Brown

Appendix: The no blame approach to state safety investigations 77
By Captain Paul Drouin

Index 81

Contributors 87

THE NAUTICAL INSTITUTE

How to use this book

This book is intended to be used in conjunction with *The Mariner's Role in Collecting Evidence – Handbook* (hereafter called the handbook), published by The Nautical Institute in 2010. The handbook gives practical guidance to those on board a vessel involved in an accident or incident and contains example lists of evidence that should be collected for the more common of these. Readers of this book will be referred to the evidence lists in the handbook as the starting point for collecting evidence on board.

There are 22 evidence lists in the handbook:

1. General, for most incidents, pages 9-12
2. Personal injury, pages 19-20
3. Illness, pages 20-21
4. Disciplinary, page 22
5. Industrial action, page 22
6. Stowaways, page 23-24
7. Distressed people, page 25
8. Diversion, page 25
9. General, for cargo incidents, pages 29-32
10. Dry cargo, pages 33-35
11. Liquid cargo, pages 36-38
12. Containers, pages 39-40
13. Pollution, pages 48-52
14. Collision and damage, pages 52-58
15. Grounding, stranding and sinking, pages 58-59
16. Salvage and general average, page 60
17. Hull and machinery damage, pages 62-64
18. Newbuilding warranty dispute, page 72
19. Port delay dispute, pages 72-73
20. Performance disputes, pages 73-74
21. Port and berth safety disputes, pages 75-76
22. Bunker dispute, pages 77-78

This book widens the scope of advice on evidence collection beyond the mariner and the Master to include those investigating the accident or incident for various reasons. Onshore evidence is also covered. The demands for evidence made by those who come on board after an accident or incident are explained, as are the demands of those who may ask for evidence from afar, possibly long after the event. The evidence lists in this book are, therefore, more detailed and specific to the purpose of the individual collecting evidence (eg surveyor, lawyer, insurer).

image: Danny Cornelissen (www.portpictures.nl)

THE NAUTICAL INSTITUTE

Chapter 1

Introduction

By John Noble, Technical Editor, *Guidelines for Collecting Maritime Evidence*

Readers of this publication will be familiar with the development of The Nautical Institute's series of books on collecting evidence, all of which have been accepted by the maritime industry as standard reference works. Dr Phil Anderson FNI led the team that produced the first book in 1989 entitled *The Master's Role in Collecting Evidence*; this was followed in 1997 by an updated book of broader scope indicated by its title *The Mariner's Role in Collecting Evidence*. This was further updated in 2006 when Dr Anderson produced the book entitled *The Mariner's Role in Collecting Evidence – in light of ISM*. In 2010, the then North of England P&I Association (now North P & I Club) and the North East England branch of The Nautical Institute combined to publish a handbook to accompany this entitled *The Mariner's Role in Collecting Evidence – Handbook*. The handbook gives practical guidance to those on board to help them collect and preserve the best evidence and accompanies this book..

All these publications dealt with the issues involved in collecting evidence after an incident on board or close to a ship, with the focus on those individuals sailing on the ship or having direct dealings with it. When such an incident occurs, however, evidence collection is not so restricted. Many different interests will want, or need, to get involved in collecting and using evidence for their own purposes and this new publication aims to reflect that. Owners, managers and mariners need to know what evidence these individuals and organisations will want and why, and how this affects their own day-to-day operation and record keeping. Those individuals and organisations when making demands on the Master, crew and shoreside offices need to be aware of others doing the same thing.

The key question for the Technical Editor was, "What is evidence?" The question may be simple enough, but the answers are many, to say the least. Some investigators or enforcers see evidence as a word only to be associated with criminal activity, while others see it in a broader light where civil action may be contemplated. Resolution of criminal matters depends on the conclusions from evidence submitted being judged to be "beyond reasonable doubt", whereas in commercial matters the burden of proof is required at the "on the balance of probabilities" level.

Clearly, the answer to this question will depend upon the mind-set of the person seeking the evidence. After an accident, a P&I Club will probably be looking for evidence that can limit its liability; a marine regulator will be looking for evidence that rules or regulations have been broken; a flag state safety investigator will be looking for evidence leading to safety lessons that can help prevent a recurrence.

Chapter 1
Guidelines for Collecting Maritime Evidence

Whoever is asking for evidence, their requirement that it covers 'who', 'when' and 'what' will be similar, as most will want to know the facts of the case. Not everyone is so interested in 'how' or 'why' an accident occurred, or what can be done to stop it happening again. Punishment can be a deterrent, but it is not always appropriate to punish individuals for failings in the system. The same piece of evidence can point both to an individual's culpability for an accident and to systemic failings that need addressing. Consequently, evidence will not always be shared among investigating bodies. This can lead to an unseemly race to grab evidence before other investigators arrive on scene. It can also lead to blocking tactics or obfuscation by those who perceive that cooperation is not in their, or their client's, best interests.

So how should the Master react after an incident when several investigating bodies are knocking on the door, all with differing requirements and priorities? Just as importantly, how can the Master, crew and management ashore be ready to produce the evidence required, much of which demonstrates day-to-day operations and compliance with the ISM Code and applicable statutes. This publication seeks to provide broad answers to those questions.

The Table of Contents demonstrates the wide range of interests that may be involved in the collection of post-incident evidence. In inviting contributions to the book, I tried, as Technical Editor, to obtain input from as many of these as possible and I'm grateful to the individuals who were willing to share their own experiences. Readers may not agree with all that has been written, and if discussion results, so much the better.

It is perhaps not surprising that there is some duplication between chapters. This serves to illustrate what is common to all evidence collection. Following an incident, however, the facts of the case will determine what specific information will be required –as will the role of the person collecting evidence.

Some chapters look briefly at the collection of electronically acquired evidence, in a general sense. There are some 27 different ECDIS systems in use on board ships. The IMO has several regulations that apply to ship-borne VDR systems and there are nearly 20 VDR systems available that comply with regulations. Generally, at present, data must be withdrawn from units by specialists, usually from the manufacturers. In addition, there are any number of publications offering guidance to practitioners who become involved in the downloading and use of electronically generated evidence.

Many ship management companies have systems where VDR information is stored remotely. Following an incident, this may need to be recovered, again by specialists. The question of who has access to this electronically generated evidence will no doubt exercise the minds of the lawyers involved. Statutory investigators will also need to rely on expert assistance to obtain the information required. Perhaps surprisingly, AIS information is more readily available, sometimes at a cost.

It is important that users of this publication are aware of the issues arising from the collection and use of electronically obtained evidence. Solutions to these issues are yet in their infancy and might even be a topic for a separate Nautical Institute publication. Certainly, as the concept of the autonomous ship becomes a reality, systems will be required to ensure accurate and reliable data is recovered and stored.

Chapter 1
Introduction

As part of the background work for this publication, help and input has been received from an expert in the field, Jennefer Tobin. She has set out here some thoughts and an outline guidance on this vital part of evidence collection.

Recording and using digital data

Systems are being developed that provide tools to a ship's crew to replace paper with digital data, captured once, at the place of work on board. For investigators, this means that electronic data will be available ashore so that a digital data set can be tailor-made to suit their needs. One example is the completion of electronic engine room logs. These give real time data to analyse and evidence of actions taken on board, such as ISM Code compliance. The system is organised to give serving mariners defined tasks so they can deliver on the ISM Code rules, as customised to their ship's requirements. These tasks are initiated at the place on the ship where the job should be performed, through a hand-held device with clear instructions on the actions to be performed. Finger entry enables the capture of digital records direct to the relevant computer system – once only capture.

This technology will have an impact on the insurance and legal sectors as it provides real time information for post-incident analysis. It will record crew performance and the effect on evidence for insurance purposes, for example, might be transformational on several fronts:

- The last dated relevant actions will provide the starting point for investigation into the causes of an incident
- Historical evidence of compliance in carrying out duties will provide context to understand how the ship was run before the incident
- This data will provide evidence that is currently unavailable
- It is easy to create additional task lists to the specification of stakeholders
- It can replace paper recording and achieve better evidence – who completed the task, when and where.

While the insurance sector is one of several that will benefit from this technology, the greatest benefit might be for serving mariners who carry the burden of responsibility and accountability.

Once the collection of electronic evidence becomes easier, there will be many issues to deal with and it may ultimately lie with the IMO to draw up some acceptable guidelines when retrieving electronic evidence from a casualty after an incident.

It is plain that in the future electronically sourced material will play an even more important part in the evidence collection portfolio. Provided the right safeguards are in place, post-incident manipulation should be impossible.

None the less, we should not lose sight of the present. The following chapters, compiled by true experts in their field, offer guidance on the breadth and depth of evidence collection.

image: ITOPF

Chapter 2

Preserving evidence on behalf of state safety investigators

The UK perspective

By Captain Andrew Moll

All flag states that are signatories to SOLAS are required to ensure that 'no blame' marine safety investigations are carried out into very serious marine casualties (VSMCs) in accordance with the IMO's Casualty Investigation Code (MSC.255(84) – Code of the International Standards and Recommended Practices for a Safety Investigation into a Marine Casualty or Marine Incident).

VSMCs are casualties involving one or more SOLAS vessels that result in total loss of a vessel, a death or severe damage to the environment. The aim of a safety investigation is to identify issues that will help prevent future marine casualties and marine incidents. While flag states shall ensure VSMCs are investigated, they are also encouraged to investigate any marine casualty or marine incident where important safety lessons can be learned, and to publish their investigation findings.

Marine safety investigation should be separate from, and independent of, any other form of investigation, and flag states have different ways of achieving this. Following the *Herald of Free Enterprise* disaster in 1987, the UK formed the Marine Accident Investigation Branch (MAIB) to formally separate the conduct of safety investigations from the business of regulating marine activity. The MAIB has 16 inspectors, almost all of them former Masters or chief engineers, who have undergone extensive training to become marine safety investigators. They are supported by a technical department staffed by professionals who are experts at piecing together electronic data to digitally recreate the circumstances of the accident.

Cooperation with other investigating organisations

An MAIB investigation will seek to establish:
- What happened
- How it happened
- Why it happened
- What can be done to prevent it happening again.

Chapter 2
Guidelines for Collecting Maritime Evidence

It will not seek to apportion blame or liability and there are, therefore, legal restrictions on the information MAIB inspectors can share with other investigating bodies. MAIB inspectors have extensive powers to board vessels, take evidence and conduct interviews. However, the key to the success of any investigation is cooperation with other investigating organisations, including P&I insurers, the local harbour authority, the local police, the marine regulator, or other states' marine safety investigators.

The Casualty Investigation Code requires investigating states to cooperate with each other as far as is practical. In addition, the MAIB has Memoranda of Understanding with the UK's Crown Prosecution Service, police forces and the marine regulator that set out how the various organisations' investigations will be coordinated and to prevent an unseemly rush to gather up the evidence before others arrive.

Preserving evidence

A team of MAIB inspectors is available to deploy rapidly to a marine casualty anywhere in the world, and their aim is to board the vessel(s) as soon as it is safe to do so in order to collect evidence while it is still fresh. However, some evidence is highly perishable. If any significant time will elapse between the accident and the MAIB's inspectors arriving on board, they will pass instructions to the Master to:

- Save VDR information and any other electronic evidence
- Photograph the accident scene
- Protect key evidence from movement or tampering
- Invite those involved to write an account of events
- Collect together any relevant documentation
- If appropriate, conduct drug and alcohol testing of those involved.

Further guidance on the evidence to be collected is given on pages 9-11.

Accident scene investigation

Once on scene, MAIB inspectors will want to establish quickly what happened. They will then plan and prioritise their investigation activities, which will depend on where the vessel is, the extent of the damage, and the future movements of the vessel and crew. If changes have to be made to the accident scene so the vessel can be moved or made safe, then the priority will be to examine the scene to the greatest extent possible before anything is changed. Otherwise, the priority is likely to be to interview those involved in the accident.

As experts in marine safety investigation, MAIB inspectors will aim to take the lead on examination of the accident scene. All MAIB inspectors undergo extensive health and safety training, and before any hazardous scene is examined in depth a full risk assessment will be conducted and appropriate risk mitigation controls put in place. The extent to which the vessel's safety management system (SMS) is functioning and can be relied upon will be taken into account before inspectors enter a disrupted or hazardous compartment and, whenever possible, inspectors will want to operate within the framework of the vessel's SMS.

By consulting with any other investigating parties, the MAIB can ensure, to the best extent possible, that all investigation objectives are met. Any testing of equipment on site will follow agreed protocols, and any items removed will be treated as evidence and handled to judicial/forensic standards. Subsequent analysis or testing of removed items will be discussed and agreed before this starts to ensure that best evidence is achieved. There are generally no restrictions on the MAIB sharing the findings of physical testing and analysis with other official organisations. When this is done, the Branch will usually request that costs are shared and separate reports are prepared if they are to be used in court.

Interviewing witnesses

Witness interviewing is vital to understanding why individuals made the decisions or took the actions they did. This understanding can be essential to helping prevent an accident recurring. However, witnesses' recollection of events can change over time as their memories become contaminated by conversation with others or they receive information from other sources. The first account of an accident that a critical witness gives is likely to be the most accurate and detailed. Secondly, once a witness has limited their account of events, possibly to protect themselves from criticism, they find it hard to deviate from that account. All this makes the early interview of witnesses by trained safety investigators potentially crucial to the investigation.

All individuals interviewed by MAIB inspectors in the course of an investigation will have their legal rights and obligations explained to them, and they will be allowed to nominate an individual to accompany them during interview. The MAIB is not allowed to disclose the names and details of those it has interviewed, nor may it disclose the answers interviewees have given.

Reports and recommendations

Once investigation at the scene of the accident is completed and all participants and witnesses have been interviewed, MAIB inspectors will widen the scope of their enquiries. They need to better understand the vessel's activities and how it is managed, the particular hazards of the trade or operating environment, and whether similar accidents have occurred in the past. At any stage, the MAIB's Chief Inspector may issue an urgent safety recommendation if this is considered necessary to improve safety.

Once the investigation is complete, a report of the investigation's findings, its analysis, conclusions and recommendations is drafted. The draft report is then circulated to individuals involved in the accident and those that have contributed to the investigation to give them the opportunity to correct factual inaccuracies and to comment on the analysis. All comments received following the consultation process are considered and, as appropriate, the report is amended before it is published.

Chapter 2
Guidelines for Collecting Maritime Evidence

The Chief Inspector is directly accountable to the Secretary of State for Transport and, because of this, MAIB reports are not scrutinised by government officials before they are published. Decisions about what and when to publish are the sole responsibility of the Chief Inspector.

Independent safety investigation reports are not always welcomed by those who wish to apportion blame or avoid liability, and the strength of a state marine safety investigation report is that its sole purpose is to try to prevent a recurrence of the accident. To do this, the reports focus on the systemic, underlying factors that combine to create the environments in which accidents happen. By focusing on these areas, the MAIB, and state marine safety investigation organisations in general, can make safety recommendations that not only affect the proximate causes of accidents at vessel or company level but also influence the rules, regulations and codes that govern the conduct of international shipping.

Evidence the Master should aim to secure

Although this guidance is provided by the MAIB, what follows describes the general principles to be followed, rather than any specific UK requirements.

> Nothing in this guidance should interfere with the emergency response, lifesaving action or efforts to preserve any part of the vessel or crew. However, once those actions have been taken and the situation stabilised, collecting and preserving evidence becomes a key priority.

Notification that an accident or incident has occurred

National regulations will generally require a vessel's Master or the senior survivor, and the vessel's owner or manager, to notify the flag state marine administration and, where appropriate, the coastal state when an accident or incident has occurred. These reports will require a description of the accident and its consequences, information about casualties and the immediate intentions for the vessel and its crew. Good quality notification information can help the flag or coastal state's safety investigators provide tailored guidance to the Master on the specific evidence that will need to be preserved ahead of their arrival and reduce the number of follow-up enquiries to be dealt with.

Keeping an accident log

Keeping a log of activity on board a ship is routine business and management of an accident or incident is, in effect, no different. The Master should aim to keep an accurate written record of the events and actions taken in the aftermath of an accident. This should be in the form of a chronological narrative including:
- Times and details of actions taken
- Internal and external communications

- Briefings and decisions
- Noteworthy events.

Perishable evidence collection

This is evidence that could be destroyed, damaged or lost if it is not captured and preserved soon after the accident. It includes evidence from drug and alcohol testing, from witnesses, of treatment of casualties, from electronic equipment and data recorders, of the accident site and physical evidence.

Drug and alcohol testing

Testing of individuals involved for alcohol and drugs should be carried out at the earliest opportunity, and the results recorded and secured. Both positive and negative results can provide important information for the accident investigation.

Witnesses

Witnesses are those directly involved in the incident. While the Master might need to establish key facts about an accident in order to ensure the immediate safety of the vessel and crew, it is preferable that witnesses are not formally interviewed before the safety investigators arrive.

Witnesses should be directed to make a written record of their recollection of events as soon as it is practical to do so. Once these written records are complete, they should be signed by the witness, collected and stored securely until they can be passed to the safety investigators. It is inevitable that most witnesses will subsequently be interviewed, but recording their recollections immediately after the accident will help ensure that the best information relevant to the accident is obtained.

Casualties

Lifesaving first aid must be the priority when injuries are suffered by any crew or passengers. The Master's accident log should record the actions taken to treat casualties and details of any drugs or medicine administered. Whenever possible, and without interfering with treatment, photographs should be taken of all injuries sustained.

Electronic evidence and data recorders

VDR

The primary purpose of VDR equipment is to provide evidence for accident investigations. There are a variety of systems, and local procedures are required to ensure that the correct actions are taken to save the VDR data in order to prevent it being written-over after the accident.

Chapter 2
Guidelines for Collecting Maritime Evidence

Closed-circuit television

CCTV systems are widely used on board to monitor cargo, machinery compartments and passenger areas, and evidence from their recordings can be invaluable for reconstructing the sequence of events and movements of people before and after an accident. Where recording capabilities exist on CCTV systems, this information should be saved and stored even if there does not appear to be any obvious link to the accident or the scene.

Electronic navigation systems

ECDIS and ECS are a crucial additional source of evidence, particularly for navigation accidents or incidents. The way the recording facilities are set up will determine how quickly the data stored is overwritten by new information. For this reason the data should be saved as soon as possible after the accident. As with VDRs, there should be local procedures to ensure the data is saved correctly, and steps taken to ensure that previous passage plans and navigational information are not deleted or altered. Photographs of the ECDIS/ECS display(s) in use should also be taken as soon as possible after an accident in order to capture the on-screen information available to the watchkeepers.

Other electronic displays

Modern vessels are equipped with a wide variety of electronic systems whose data can assist a safety investigation, including NAVTEX, course recorders, data loggers and control panels. As soon as possible after the accident, records should be made of relevant settings, the displays photographed and, where possible, the data on the systems should be saved.

Photographing the accident site

When the ship or the cargo have been damaged in an accident on board, a photographic record of the accident scene should be made before it is altered. If equipment has to be moved or the accident site altered in any way, photographs should be taken both before and after the changes.

When photographing an accident site or consequential damage, these principles should be followed:

- Start by taking some wide-angle images of the accident scene in order to provide context for the subsequent, more detailed images. If possible, include well-defined, measurable objects such as a door or the full height of a compartment to give the viewer a sense of scale.

- Take images of associated equipment or systems, also to offer context to the accident site. For instance, where equipment has failed, all the surrounding or connected systems should be photographed; include gauge readings, system pressures and valve positions. If there is a duplicate of a failed system on board, take photographs of this system. This will enable investigators to undertake a comparative analysis between a failed and working system in the same vessel.

- Take images of the accident scene from as many angles as possible and increasing levels of detail. The first images should be taken with light conditions as similar as possible to the time of the accident. For subsequent images, use additional light sources to illuminate darker areas.

- Make a written record of the sequence of photographs taken. This will assist investigators in their subsequent analysis of the imagery. Also, make a sketch of the scene, labelling key components and items of interest.

Physical evidence

Physical evidence is equipment or other objects that have, in some way, been involved in the accident, such as failed engineering components, parted ropes or wires. Once a full photographic record of the accident scene has been made, physical evidence should be preserved. Under no circumstances should any damaged or broken equipment be discarded or otherwise disposed of. All physical evidence should be preserved for the safety investigation.

If physical evidence cannot be stored or preserved, it should be recorded to the best extent possible through photography, sketches and measurements. Equally, damaged equipment or machinery should not be examined or dismantled unless this is essential for the safety of the vessel. Ensure that any items removed are securely packaged to preserve them for later examination by safety investigators.

Non-perishable evidence

Once the collection of perishable evidence is complete, other evidence that will be required to support a safety investigation should be gathered. Documents can be photocopied, photographed or saved onto an electronic storage device such as a memory stick. Under no circumstances should any vessel documents be altered, amended or destroyed after an accident.

An outline of the documentary evidence is given on pages 9-12 of the accompanying handbook, *The Mariner's Role in Collecting Evidence – Handbook*.

Safety investigators are likely to provide more detailed requirements once the nature of the accident is known.

image: Danny Cornelissen (www.portpictures.nl)

THE NAUTICAL INSTITUTE

Chapter 3

The Master's responsibilities for collecting evidence

By Captain Ian Odd

Once an accident, or incident, concerning their vessel has been stabilised, Masters quickly turn their minds to the reporting and investigation in which they will now be involved. This might be fairly straightforward in the case of a minor personal injury or exceedingly complex in the event of a more serious injury or damage to property. Whatever the case, accurate reporting to owners or managers is essential as the possibility of enquiries, investigation or even litigation in the future must be considered. Form-filling will be needed, as well as the gathering of evidence by interviewing people involved, including witnesses. If the incident was caused by mechanical or material failure, it will be necessary to secure the damaged item for future expert assessment.

Checklists are useful tools whenever an incident has occurred. They serve to remind us what we need to do. It is very easy to forget to record or to take action regarding something vital that might be difficult to accomplish later. A comprehensive series of checklists is available in the accompanying handbook, *The Mariner's Role in Collecting Evidence – Handbook* (hereafter referred to as the handbook).

Whether you fill out a checklist or make a report, depending on the demands of your company, there are important areas to include that need to be recorded accurately.

Date, time and exact location

Always state if the time used is local or UTC. If local time is used, then note the time difference between local and UTC. The date should always be written clearly, eg 21 November 2017.

If at sea, always use latitude and longitude for the position. If the vessel is in port or port approaches, state exactly the vessel's position within this area by geographical means.

Actions taken

What was done by the Master or other crew members immediately following the accident or incident. Note the time the Master was made aware of the incident.

Chapter 3
Guidelines for Collecting Maritime Evidence

Secure evidence

An incident on board ship causing injury to the ship's crew or shore workers might involve the failure of some part of the ship's equipment, such as a broken wire, shackle or anything else involved in lifting. Whenever possible the damaged part should be taken out of service, labelled and kept in the care of the Master. Clearly this will not be possible with larger items. Always be aware that people could try to cover up their possible negligence by hiding or destroying physical evidence.

In the case of a navigation incident involving the bridge team, always ensure that the VDR data is secured before being overwritten or destroyed. Many VDRs will have a simple means of securing data. Do not rely on being able to use unsaved data at a later date. Only the Master should be able to access VDR data. Charts, nautical publications, night orders, Master/pilot exchange information, passage plans, checklists and any other items in use at the time must be secured to prevent illegal changes or destruction of evidence.

On vessels fitted with an electronic chart system, ensure that the records of past positions are not deleted. Log books and bell books should be secured immediately after the incident to prevent data being erased or pages destroyed. Some ships might use VHF logs recording when VHF transmissions were made or received and the brief message content. Should an incident occur when manoeuvring in port, note the names of the pilot and any tugs used.

Persons involved, including witnesses

The names and ranks of all crew involved in the incident and any witnesses must be carefully recorded. In the case of people who are not ship's crew, it is very important to record their full names, job title and home addresses. If individuals are not willing to give their full details, this fact should be noted. An up-to-date crew list should be filed with the incident documentation.

Photographic evidence

Digital photography makes it very easy to record evidence after an incident, and sometimes before an incident, in the case of CCTV footage. Many people now carry smartphones that can be used for taking good quality still or video pictures. It might be possible to secure CCTV footage and prevent it being overwritten. Any digital photos or video footage should be backed up to prevent loss or deletion. Witnesses should be requested to provide copies of any photographic evidence they might have. The Master should try to take good quality pictures illustrating the immediate and surrounding areas where the incident occurred. The camera's macro function should be used for better close-up pictures. The Master has the right to restrict people from taking photographs after an incident.

First Aid treatment given

If someone has been injured in an incident they will have been treated on board, unless in port where paramedics or ambulance crew have attended. It is important that a record is kept of treatment given and by whom. A photo of the casualty before and after treatment will be useful but bear in mind that a casualty should always be asked for permission, if possible, before taking photographs. A careful log should be kept of any medical advice received from shore.

Advise owners/managers

It is important that the vessel's owners/managers are advised of the incident as soon as practically possible. Timing will depend on the seriousness of the event. Record the date and time of any messages or phone calls and compile a written summary of any phone or VHF calls made.

Advise local authorities (if applicable)

The local port state and port authorities should be advised of the ship's status or injuries to people on board as soon as possible. The date and time of the call needs to be carefully recorded and a written summary made of it.

Logging of incident

A lot can happen in a short space of time and the ship's log will not be written up immediately. However, all rough notes taken after the incident will count as evidence and must be preserved. Logging does not just involve the ship's log book. Anything written down in bell books, on checklists or any other notebook involved in the incident is a log of events. Similarly, bunker sounding records, ballast tank records and any other stability information will need to be preserved.

Some Masters keep diaries. They should limit anything they record to the known facts and not record opinions as their diaries might be called into evidence. All logs should be signed by the OOW and the Master. Do not cross out anything and certainly no correction fluid should be used. Anything written in error should have a single line struck through it so that it is still legible. The corrected text should be inserted nearby and the entry initialled and dated by the writer. The Master should look for any evidence that a page might have been removed. It might be worthwhile to check waste paper receptacles.

Interviewing witnesses and obtaining written statements

Few Masters are experts at interviewing people following an accident, so here are some essential guidelines. It is important that any interviews are conducted as soon as possible

Chapter 3
Guidelines for Collecting Maritime Evidence

after the incident while events are still fresh in people's minds. Before interviewing anyone it is essential to try to put people at their ease. They should be told that the intention of the interview is to get to the truth and not to apportion blame. The Master should advise them that what they say is being recorded either in audio or written format. A written statement should be made up from the notes or audio recording and subsequently shown to the interviewee for them to approve and sign. They should always be given a copy of their signed statement.

It might be necessary to try to segregate witnesses or those involved in an incident. This is to try to prevent undue influence of interviewees. It can be very difficult to get to the truth. Never assume that the interviewee is telling the whole truth. There are many things that could affect what the interviewer is told, for instance it is common for people to be concerned that what they say may get someone else into trouble. It is also not unusual for a person to deliberately try to incriminate someone else. Inter-departmental rivalry is another factor to be considered in the collection of factual evidence.

In the case of major incidents that would require an inquiry by state safety investigators, see the advice in Chapter 2.

Weather at time of incident

It is essential to log the weather conditions at the time of the incident as this can be an important factor in investigating the root cause. Many an accident can be put down to a vessel in heavy weather, shipping seas and rolling or pitching heavily. Precipitation in the form of rain, ice or snow can make decks slippery. Excessively hot or cold temperatures can also be a cause. For navigation incidents, any conditions affecting the state of visibility may be critical.

Lighting

This can be critical. Note whether it was day or night. Was the area sufficiently illuminated for the persons involved to carry out their tasks safely? If artificial lighting was in use, what type was it? Was any portable lighting in good condition and safe to use?

Equipment in use

Any equipment or tools in use at the time of the incident should be carefully checked to ensure they are in a serviceable condition. Any maintenance carried out on the equipment should be recorded. Checks should be made to ensure that no unauthorised modifications have been made to tools or equipment. In the case of electrically-driven items, the status of any safety cut-outs or trips should be ascertained. It is not unusual to find that a faulty cut-out device has been bypassed instead of replaced.

Chapter 3
The Master's responsibilities for collecting evidence

Permits to work and risk assessments

Certain jobs require a written permit to work to be completed before starting a job. Examples might be hot work, enclosed space entry, working at height or working with pressure systems. In each case a written risk assessment should have been completed. Jobs not requiring any formal permit or risk assessment should still have been discussed at an informal 'toolbox' meeting at the start of the working day or before a job starts. The officer in charge should conduct a meeting with crew involved at the start of the working day. Records of permits, risk assessments and toolbox meetings should be collected as important evidence where applicable.

Training and qualifications

Lack of proper training is a frequent cause of accidents. There should be records on board of all collective training carried out, including drills. Individual training records should also be available. Many companies also include computer-based training (CBT) for their crews. Individuals should carry records of all the shore-based training they have undertaken. Where relevant, copies of any training records should be retained as part of any investigation. It might be necessary to make copies of various crew and officer qualifications in addition to recording them.

Warning signs

Today's commercial vessels have a multitude of warning signs and posters in various locations on board, particularly in hazardous areas. Photographic evidence should include what signs were posted in the vicinity of the incident, if appropriate.

Incident log

This should be in the form of a statement of facts in chronological order. Many things could have happened at the same time but as much as possible should be included in the correct order. Timings should be as accurate as possible. No comments, opinions or anything else of that nature should be included.

Filling out an accident or incident log

All shipping companies should have an accident reporting form as part of their Safety Management System (SMS). Some companies might have two different formats, one for P&I incidents and the other for hull and machinery (H&M). The former is to be used for all incidents involving injury to crew and damage to third party property. A pollution incident would be recorded on this form. An example of an accident reporting form for P&I incidents is given on page18.

Chapter 3
Guidelines for Collecting Maritime Evidence

Accident Report (P&I case)

Accident Report No.: ACR001	**Date of issue:** 23 November 2017
Claim no (office use only):	**Voyage no:** 18
Ship's name: Nonsuch	**To:** Jakarta
Casualty location: Jakarta	**Casualty date:** 21 November 2017
Lat.: 006° 06'S **Long.:** 106° 54'E	**Time (Local Time):** 1530hrs

Casualty details
AB Smith had been given the job of painting the forward mast. As he was being hauled aloft, the gantline parted and he fell approximately 2m to the deck. This caused him to suffer a broken left forearm. He received First Aid and his broken arm was put in a sling. He was given painkillers and seen by a doctor soon afterwards.

Likely causes of incident

Immediate cause
The fibre rope gantline parted. Upon examination it was found that the rope had been stored near some paint thinners which had leaked out and damaged the rope fibres.

Root cause
Improper stowage of rope and paint thinners and failure to properly inspect rope before use. This will also constitute a non-conformity of the vessel's SMS.

Corrective actions for preventing recurrence

The Master held a safety meeting of all the crew to explain how the accident occurred and how it could be prevented. The Master instructed senior officers to ensure items are correctly stowed and inspected thoroughly before being used in future.

Schedule for completing corrective action (circle one): (One week)   One month   Three months
Corrective action implemented and date of completing the action: 23 November 2017
Case closed on: 23 November 2017

Reported to Club:		**Reported to Owners:** 21 November 2017	
Date:	**Email:** crew@Easternship.com		**Fax:** N/A
Local agents name: Eastern Shipping Agency	**Hospitalisation needed:** Yes / (No)		
Date crew disembarked from vessel: 22 November 2017			**Port:** Jakarta
Date crew sent home:	**Airticket arranged with:**		
Replacement crew:	**When:**	**Where:**	
Local agents for replacement crew:	**Additional cost incurred:**		
Injured crew ID:			
Master's signature (always):			

Chapter 3
The Master's responsibilities for collecting evidence

Notes on the example

- The Master should fill in all the boxes completed in this example. The boxes left blank are for owners
- Send a follow-up report later if it is not possible to complete the report in one go. Add extra pages and relevant documentation if it is not possible to include all the information on a single form
- Report numbering should follow the ship's system
- The date of issue is the date of initially filling in the form. It may not be the same as the date of the incident
- Give as much information as possible about exactly what happened
- The case closed date will generally be the date the owners decide in consultation with the Master
- No physical evidence should be disposed of without the owners' permission
- The Master should sign the form when complete and give the ID of injured crew, if appropriate.

The Master's report is made after carefully considering the evidence. In this example, this would ensure the damaged rope was safely stowed away until the owners instruct how to dispose of it. Photos of the scene and the casualty would be made. The Master will have found out why the rope was not inspected before being put into use and why it was stowed where it could become damaged and made unsafe for use. When the rope was first brought into use and when it was last properly examined would be established.

Any supporting documented evidence, including the manufacturer's certificate and any subsequent test certificates, should be attached to the report. Selected photos can be printed off with a reference to where all digital photographic and other evidence is stored on the computer. Alternatively, the photos could be burned onto a disk.

H&M covers damage that would normally be covered by the vessel's hull policies and which is not covered under P&I rules. An example of an accident reporting form is given on page 20.

Chapter 3
Guidelines for Collecting Maritime Evidence

Accident Report (damage case)

Accident Report No.:		Date of issue:
Claim no (office use only):		Voyage no:
Ship's name:		Insurer:
On voyage from:	To/lying at:	With cargo: Y / N
Vessel's approx. position when damage occured or discovered – Lat.:		Long.:
Casualty date:		Time (Local Time):

Damage to vessel. Extent of damage

Likely cause of damage or reason for discovering same

Immediate cause

Root cause

Damage reported for survey:		Reported to Owners:	
Date:	Email:		Fax:
Survey made by:		Survey made of:	
Damage repaired by.			
Own damage:		3rd party damage:	
Repairs are: (when damage to machinery, state maker and type of engine)			
Schedule for completing repair (circle one):		One week One month Three months	
Corrective action (repair) implemented to prevent recurrence and date of completing the action:			
Case closed on:			
Master's signature (always):		Chief engineer's signature (machinery damage):	

THE NAUTICAL INSTITUTE

Chapter 3
The Master's responsibilities for collecting evidence

The accident report (damage case) will be filled out in much the same way. This will be done in close consultation with the chief engineer who will sign the document where machinery damage is involved. Again, wherever possible, the damaged items should be retained until the owner issues disposal instructions.

All reports made should be factual without any speculative or suggestive comments. Remember that a Master's private diaries could be used in evidence if a case should become a subject of litigation. Careful filing of all written and digital matter regarding the incident is essential. Masters should retain their own copies as sometimes an incident will be referred to many years later. This will help recall as Masters will be expected to answer questions about incidents, even minor ones, at that stage.

Dealing with officials coming on board

Surveyors, inspectors and officials can give conflicting instructions. The Master should be advised by owners who to expect to board the vessel and whether they are appointed by the owner or third parties. No evidence, either physical (such as damaged items, including cargo or bunker samples) or records (including logs and witness statements), should be handed over to anyone without the express permission of the owners. A complete record of what evidence has been handed over should be kept by the Master. This record should include not only to whom it was handed and when but also who gave permission to hand it over.

For more information on this see *Casualty Management Guidelines*, published by The Nautical Institute.

Dealing with the media and social media

Queries from the media should be referred to owners or managers. Crew should be instructed not to talk to anyone from the media and to advise the Master of any such approaches. Masters must be aware that unscrupulous individuals may misrepresent themselves either by phone or in person.

Crew should also be reminded that images or comments posted on social media can become public around the world very rapidly, driving media interest and uninformed speculation.

For more information on this see *The Nautical Institute on Command*, third edition and *Casualty Management Guidelines*, both published by The Nautical Institute.

image: A R Brink & Associates

Chapter 4

The P&I approach

By Chris Adams

As an insurer of the liabilities of shipowners, operators or charterers, a P&I Club has a keen interest in ascertaining the facts and collecting all relevant evidence. Without this evidence, it cannot effectively investigate the circumstances of any incident giving rise to liabilities falling within the scope of its cover. Reliable evidence is an essential prerequisite to determining whether, and to what extent, a Club's member is exposed to liability in relation to any particular incident. The Master, officers and seafarers on board any vessel involved in any casualty likely to give rise to a claim upon the owner's P&I cover must understand the importance of full and frank disclosure of the facts relating to the incident.

The role of the P&I Club, through its appointed representatives who may attend a vessel to investigate a casualty, is solely to determine the facts. The Club has no interest in attributing blame to individuals for any incident that may have occurred. As underwriters of liabilities, the Club recognises only too well that humans are fallible and that mistakes will be made. The Club is there to handle the consequences. However, for those consequences to be handled effectively and the resulting liabilities minimised, it is vital that shipboard witnesses understand that the owner's P&I Club is on their side and that they cooperate fully so that reliable evidence can be collected.

Where a casualty has resulted from an individual's error, there can be an understandable reluctance to fully recognise and disclose what has occurred because of concern about possible consequences. Such an approach is always unhelpful. Events giving rise to casualties are always investigated rigorously and inconsistencies in evidence will always be uncovered eventually. It is always preferable, and far less costly, for there to be no such inconsistencies in the evidence.

If there are any inconsistencies, it is better these should be determined and resolved by the owner's P&I Club's investigative team at an early stage rather than being revealed by opposing lawyers in any subsequent litigation. It is only by knowing the truth and the full facts – no matter how unpalatable some of those facts might be for individuals – that the P&I Club can properly and effectively defend its member's interests.

Establishing the facts

Precisely what evidence is required by the P&I Club to protect its member's interests varies from case to case. The core requirement, however, is for any and all information and documents that will assist in establishing the underlying facts giving rise to the incident.

Chapter 4
Guidelines for Collecting Maritime Evidence

That information will then be used to determine whether the member is liable for the incident or if the evidence can be used to establish a defence – partial or otherwise.

The Club must be notified promptly of any incident that is likely to result in a liability or liabilities arising that fall within the scope of its cover. Prompt notification will enable investigations to be undertaken as quickly as possible and for all appropriate evidence to be collected before that evidence either dissipates or is otherwise lost. It also provides the opportunity for witness statements to be recorded before recollections fade.

Vessels should have on board a copy of the ship's certificate of entry from the owner's P&I Club. That will identify the P&I Club concerned. There should also be on board a copy of the P&I Club's Rule Book and List of Correspondents. Alternatively, details of the Club's listed correspondents can be found on its website. The Club's correspondents are generally independent firms that specialise in investigating P&I claims on behalf of their Club principals. They are not agents of the Club, but firms that the Club recommends to its members as part of the claims handling process. When they act, they do so on behalf of the Club's member.

The Club correspondent network, which extends worldwide, enables any Club to very quickly arrange for an appropriately qualified and experienced individual to attend the vessel to provide assistance to the Master in dealing with the immediate aftermath of any incident. That representative will also investigate the circumstances of the incident and collect all available evidence that will help determine the cause and the assessment of liability. Firms that are part of the correspondent network are frequently the same whether the vessel's P&I insurance is provided by a Club which is a member of the International Group of P&I Clubs, or by one of the alternative providers of such insurance.

Lawyers and commercial correspondents

Firms that are used as correspondents fall into two broad categories – either lawyers or so-called commercial correspondents. Which will be appointed depends on the nature and severity of the incident – more serious casualties obviously require a more extensive investigative team. Routine incidents, such as minor damages to cargo, may be investigated solely by a representative of the correspondent firm – possibly assisted by a surveyor who might be from the same or another firm.

More serious incidents (such as collisions, damage to fixed or floating objects, loss of life or personal injury or oil pollution) will require a more extensive team, possibly involving maritime lawyers and experts in the subject matter of the claim likely to result from the incident. For example, consulting civil engineers may be instructed in the case of extensive damage to port structures; specific commodity experts will be used when there is extensive damage to cargoes of that commodity. Some subject matter experts may have to be mobilised from elsewhere and may not arrive on scene for several days. The role of these experts is principally to assist in minimising the extent of the loss or damage.

It may also be necessary for lawyers to attend to record evidence. Sometimes claims or disputes arise and an amicable settlement cannot be reached, so litigation or arbitration in another country is undertaken. In many instances, English law and the jurisdiction of the English courts or an arbitration tribunal may be stipulated in any applicable contract, or may be agreed as the dispute resolution mechanism by the parties involved. For such cases, it will be necessary to collect evidence in the form appropriate to the applicable procedure in case it should be necessary to have the issue of liability actually determined by a court or arbitration hearing.

Many of the individuals that a Club may instruct to attend a vessel for the purpose of gathering evidence may themselves be former mariners, as indeed are many of the Club's claims handling staff. The experience that ex-mariners bring to the handling of the consequences of a casualty, its investigation and the gathering of evidence is frequently invaluable. An ex-mariner's insight can assist in quickly identifying any technical shortcoming or inconsistencies in evidence. That insight will also help to ensure that the evidence gathering process is thorough, effective and serves to best protect the interests of both the member and the Club.

After a casualty

Core tasks of those attending a vessel:
- Establishing immediate action necessary to contain the situation and prevent escalation
- Determining a strategy for handling claims
- Investigating the circumstances of the incident
- Collecting all relevant evidence.

This will enable the results of the investigation and the evidence collected to determine the cause and the extent of the loss, determine where liability lies, and to negotiate the most favourable outcome for the member and the Club.

There may be instances where there is unavoidable delay between an incident and the arrival of the investigative team. This may be because the vessel is at sea and several days away from port. Alternatively, the vessel may be at a remote place where there is no local Club representation and a correspondent may have to travel from the nearest available location.

Preserving evidence

Before the arrival of those instructed to assist and investigate, it is vitally important that mariners on the vessel take steps to preserve all evidence that will ultimately be of assistance to the Club's appointed team. This will include all relevant documentation and any physical evidence. Photographic evidence of the site of the incident taken in the immediate aftermath is particularly valuable. There is generally no shortage on vessels of devices able to take digital images. One can never have too many photographs. To maximise the value of such material, these should be taken in good light from as many different angles and ranges as possible.

Chapter 4
Guidelines for Collecting Maritime Evidence

When personnel begin to arrive at the vessel, everyone needs to be aware that there may be more than one P&I insurer involved with a particular ship. If the vessel is operating under a charter, there may be one or more charterers who might also have an interest in any particular incident, especially if it is cargo-related. Each of those charterers is likely to have their own P&I insurance for charterers' liabilities and each may wish to conduct its own investigation into the incident.

Things can be very complex. If great care is not exercised, there is a risk that Masters, officers and crew might provide information, documents or other evidence to someone who may not be entitled to receive it. For example, a surveyor might arrive on board and simply announce that he has been instructed to attend on behalf of the P&I Club. The unwary Master and officers might assume that he represents their owner's P&I Club, whereas the individual might actually be representing the charterers whose interests are not the same as those of the owners.

If there is no clear understanding of the identity of that surveyor's principals, information and documents may be provided that might in fact be protected by legal privilege (see page 41 for more details), or make unguarded comments that could be prejudicial to the vessel owner's defence. It is therefore imperative that checks are made to determine who any particular individual is representing before any information is given.

One of the most important categories of evidence that will be collected, particularly on a casualty of any severity, is witness testimony in the form of written statements from people who saw what occurred and are in a position to recount those events. This will involve being interviewed and questioned about the circumstances of the incident by the lawyers appointed on behalf of the vessel's owners. The lawyers will then incorporate what has been said into a written statement that will be provided to the individual witness who should read it carefully to check that the statement is factually accurate. Any inaccuracies should be identified and corrected and the statement should only be signed by the witness when they are satisfied that it is completely accurate.

Hopefully, the collection of comprehensive and reliable evidence will enable claims arising from incidents to be resolved amicably. However, this is not always possible and it may be necessary for the dispute to be resolved in a court of law or by an arbitration tribunal. In this case it may be necessary for the witnesses of fact to attend to give evidence and for the reliability of that evidence to be tested by cross-examination. This may occur many months or indeed years after the event and the written statement given at the time will be a key point of reference.

Reliability of witness evidence

Unfortunately, the ultimate outcome of litigation may depend very heavily upon the reliability of witness evidence. This can be problematic. In addition to the human fallibility that often results in marine casualties, humans are often unreliable witnesses. The human memory is unreliable and individuals can easily miss things they are not expecting to see.

Chapter 4
The P&I approach

A search on the YouTube website using the words "observation test" will identify a number of video clips that admirably illustrate the fallibility of the human memory. The accuracy of recollection also diminishes with the passage of time, so it is vitally important for witness testimony to be formally recorded as soon after an incident as possible.

An obvious way to aid recollection is for witnesses to make notes about any incident as soon as possible after it happened. However, great care is needed. If the incident is likely to result in litigation, that in turn will involve 'disclosure or discovery'. This is a process whereby all relevant information or documents are disclosed and provided to the other side in the litigation. This would include any written notes prepared by witnesses if they are not protected by legal privilege (see page 41).

There can never be any certainty that such written notes are factually accurate, and any inaccuracy that may be present might arise perfectly innocently. Ask any two individuals to watch an event and then describe it, and there will inevitably be differences in perception. Such differences need to be resolved by examination of the evidence as a whole in the evidence gathering exercise. Further, notes made by witnesses may express opinions about the incident and why it occurred that may not necessarily have been formed on consideration of the complete picture.

Legal privilege is the principle that provides protection from disclosure of any material that is produced in contemplation of litigation. This is to enable witnesses and others to communicate fully and frankly with the owner's lawyers without fear that what is said or provided will necessarily have to be handed over in the disclosure process. This then enables the evidence to be gathered and presented in a manner that is completely consistent with known and immutable facts.

In order to ensure that there is a good prospect of successfully claiming legal privilege over any witness notes that may have been made, these should ideally always be addressed to the owner's lawyers and contain the statement that they have been prepared in contemplation of litigation. For further information on this see Chapter 6.

Case study 1

A vessel fails to turn as expected during a manoeuvre and comes into contact with another vessel in the vicinity that is also underway. The OOW might jump to the conclusion that the failure to turn was due to a malfunction of the steering gear and make a note to that effect. However, that supposition has not been subject to critical examination or verified in any way. It may well be that the steering gear was operating entirely satisfactorily and that the failure to turn was attributable to other factors.

The existence and potential disclosure of a note referring to a machinery malfunction would clearly introduce an unnecessary complication, which at best would take time and cost to overcome, or at worst could damage the owner's defence.

Chapter 4
Guidelines for Collecting Maritime Evidence

The susceptibility of witness statements to error and subjectivity makes evidence that is derived from incontrovertible sources preferable; the reliability of witness testimony will always be tested by comparison with such material. In cases of collision, before the advent of AIS, the reconstruction of events leading up to a collision was frequently a time-consuming and costly process. It involved testing of witness evidence against more reliable data – such as automated data sources like engine movement loggers, course recorders, VTS data, VHF transcripts and the results of a speed and angle of blow survey following inspection of the physical damage to the ships. Considerable effort was frequently required to reconcile subjective witness testimony with the harder objective data, such that the events described could be seen 'to work' in reconstructing the events leading to the collision.

Now the task is much easier as the availability of AIS records enables the tracks of vessels involved in a collision to be reconstructed within a few hours of the casualty having occurred. That material is frequently available on public source websites such as YouTube, enabling a very early assessment to be made of the likely allocation of liability by anyone with a familiarity with the Collision Regulations (Colregs). In addition, the data downloaded from the vessel's VDR also helps to accurately determine the facts of any navigation incident.

In some casualties it may be necessary to rely extensively upon witness evidence – for example, where the vessel may be lost and there is no opportunity for other evidence to be either preserved or recovered from the VDR. In those cases, witness evidence will be subject to the closest scrutiny.

Case study 2

The case of the *Atlantik Confidence* [2016] EWHC 2412 (Admiralty), involved the foundering and total loss of the vessel and its cargo in the Gulf of Aden in March 2013 following an engine room fire. Because of the lack of physical evidence, the determination of the cause of the vessel's loss, and whether the owners were entitled to a decree limiting their liability, depended very heavily upon the evidence of witnesses from the vessel and the owning company and from expert witnesses. The judgment in this case demonstrates the extent of the scrutiny and evaluation of witness evidence that was undertaken by the presiding judge Mr Justice Teare.

Such cases are thankfully rare, but no matter how serious an incident might be, comprehensive and reliable evidence is always required to protect the interests of those who need to rely upon it. Mariners have a vital role to play in the evidence gathering process.

Chapter 5

The surveyor's perspective

By Captain Sanjay Bhasin

The investigation of any incident depends on the quality of evidence collected. This, in turn, depends on the time elapsed since the incident, the geographical location of the incident and the experience and qualifications of the investigator.

A major incident or casualty is classed as one which involves personal injury or fatalities, or significant damage to a vessel, its cargo, to a shore facility or pollution of the marine environment. The collection of evidence is an essential process after any such event. The evidence collected by the surveyor will be critical in the reconstruction of the incident and determination of the cause. The results are likely to have legal and financial implications for one or more parties involved, including crews, stevedores, shore personnel, owners, charterers, cargo interests, port authorities and port owners/operators. Inevitably, the insurers for the various parties will also be affected.

Following a major incident, a marine surveyor will almost certainly be appointed by the owners, their P&I Club or hull and machinery (H&M) insurers to collect relevant evidence and investigate the cause. Depending on the type of incident, charterers, cargo interests, stevedores and port operators may also appoint a surveyor to attend.

Legal implications and procedures may prevent complete access to the vessel's crew and documents for surveyors appointed by parties other than the shipowner. Even in cases where attendance on board is permitted, owners may impose restrictions on the type of access that any surveyor may have. The surveyor appointed on behalf of charterers or cargo interests should be provided with information regarding limitation of access to the vessel, crew and cargo, which should be strictly followed.

The evidence that can be obtained on board includes visual, documentary and first-hand personal evidence from crew who witnessed the incident. Evidence of procedures and practices on board, and those of the company, will also be required to verify the vessel's compliance with the ISM Code and applicable statutes. It is important that the appointed surveyor is experienced with the type of incident that has occurred and the evidence that needs to be collected.

A complete investigation will require analysis of the evidence obtained, and provision of opinions based on that evidence. This is secondary and will take a considerable amount of time after the initial attendance. In the process of collection of evidence, a surveyor should not focus on who, or what, is to be blamed.

Chapter 5
Guidelines for Collecting Maritime Evidence

Depending on the nature and type of incident, evidence collection may not be limited to the vessel(s) involved. The surveyor may need to seek evidence from the port authority, pilots, local surveyors, port and berth operators, berthing staff, stevedores, agents, eye witnesses ashore, weather services and external data retention services. Evidence from vessels in the vicinity may also be relevant. Priority should be given to what is immediately available, especially if it is likely to be otherwise lost.

It is generally difficult to obtain evidence from port authorities or pilots. In some areas of the world it is almost impossible. However, attempts must be made to obtain information and the responses received should be documented.

The collection of evidence after an incident includes:
- Relevant documents, electronic data and information from the vessel's staff
- Inspection of equipment that was in use on board
- Interviews and statements from the Master and others involved. Formal statements are normally taken by lawyers but they may use the surveyor's marine expertise when interviewing the Master and crew
- Information on port operations, port layout and facilities, pilotage and other port parameters that may be relevant to the causation
- Evidence of the environment in which the incident occurred from the vessel and weather services
- Historical data of similar incidents in the area, if applicable
- Positional data from external service providers.

It is likely that the principals who appoint the surveyor will also want evidence for the assessment of damage and other liabilities. This chapter deals only with evidence collection related to cause of the incident and gives the example of navigation incidents.

Navigation incidents

Investigation of navigation incidents requires the expertise of Master mariner surveyors, preferably with command or pilotage experience. If engine or steering failure played a role in the cause of the accident then a suitably qualified and experienced chief engineer should be engaged. Detailed maintenance, overhaul and repair records of machinery under the vessel's planned maintenance system, will also be required in those cases. Such detailed scenarios are outside the scope of this chapter, which covers the following navigation incidents:
- Collisions
- Groundings
- Contact with fixed or floating objects (FFO)
- Wash damage.

Both general evidence relevant to navigation incidents and evidence for the specific type of casualty should be collected. The documents and evidence mentioned in this chapter are not exhaustive and additional evidence may be needed, depending on the

Chapter 5
The surveyor's perspective

type of vessel, incident and location. The suggestions for evidence given in this chapter should be used in conjunction with the lists on pages 52-59 of *The Mariner's Guide to Collecting Evidence – Handbook*.

Ideally, collection of evidence should be undertaken as soon as possible after the incident. As the vessel involved is very likely to be fitted with a VDR, the first step would be to confirm whether data had been saved before the overwrite period. VDR provides invaluable contemporaneous evidence and greatly assists in the reconstruction of an incident and in the investigation.

If the VDR data has been saved, the surveyor should obtain a copy of the saved data, together with the playback software. It may not be possible to playback the data immediately, as a technician from the manufacturer may be required to extract it. If the VDR playback is available, this can greatly assist in taking statements from the Master and other crew involved in the incident. Real-time playback helps the interviewee when recalling actions taken before the incident and the reasoning behind them.

While VDR data has made collection of evidence and investigation potentially simpler, it has also raised the level of investigation because bridge team conversations are now available, and the discussions (or lack of them) and radio and telephone communications before and at the time of the incident also play a major role in a reconstruction.

It is also possible to obtain evidence of a vessel's positions, courses and speeds using AIS data. There are companies worldwide which can supply this information. In the absence of any VDR data or a delay in setting up the playback, surveyors should, with approval from their principals, enquire whether AIS data relating to the vessel for the time and place of the incident is available. AIS data is public and can also be obtained by other interested parties. The small costs involved make it worthwhile to obtain it at the outset.

For navigation incidents, it is essential to speak to the Master first and obtain the basic facts. That should be followed by more formal discussions with the Master and the bridge team to ascertain what happened. If a lawyer is present, then it is likely that statements are being, or have been, taken.

It is necessary to inspect the bridge and the navigational equipment. Original documents are generally retained by lawyers or owners, a list made and a receipt given. The surveyor must also take photocopies of all the documents, irrespective of the fact that the originals will be retained and made available later. If there is any difficulty in photocopying, the documents can easily be photographed using digital cameras. If the surveyor is required to collect original documents, great care must be taken for their security and prevention of any damage to them. The originals are likely to be inspected later by other parties if the case goes to litigation.

When on board, surveyors must be suitably equipped with safety gear and only proceed to areas when escorted by the crew. It is important that no bridge or other equipment is operated by surveyors on their own, and the officer responsible should be asked if it is necessary to check its operation.

Chapter 5
Guidelines for Collecting Maritime Evidence

Investigation of navigation incidents requires experience or familiarity with the bridge equipment. However, given the large number of manufacturers and models available, it may not be possible to ascertain the features of a particular model of equipment that is fitted on board. Surveyors should consult the onboard operator's manuals and make photocopies of relevant sections for analysis later to see if the equipment was set up correctly.

General evidence to collect for navigation incidents

- VDR download and playback software
- Vessel's particulars
- Crew list
- Bridge team and their qualifications
- Crew involved and where stationed
- Vessel's navigation equipment on board
- Equipment inoperative or not used at the time of the incident
- Settings of equipment in use
- State of engines and steering motors in use
- Vessel's draughts, list/heel at the material time
- Manoeuvring diagram
- Vessel's statutory certificates
- GA and capacity plan
- Deck and engine logbooks, official and working, if more than one is kept
- Bell book for deck and engine room
- Engine data logger
- Soundings book
- Ambient conditions (wind, swell, current and visibility)
- Navtex reports
- EGC reports
- ECDIS history
- GPS, GPS tracker
- Working chart
- Chart correction log
- Passage plan
- Course recorder trace (important if VDR data has not been saved)
- Echosounder trace and history if electronic
- SMS checklists for bridge procedures and watchkeeping
- Chapters in the SMS manuals relating to navigation
- Chapters in the SMS manuals relating to role of pilots and Master/pilot information exchange if on board at the time of the incident
- Master's report of the incident as required by the vessel's SMS
- Any communications with owners, charterers, port state control or other third party
- Note of protest, if any
- Antenna plan

- Work and rest hours of the bridge team and others involved
- Results of alcohol and drug tests, if carried out after the incident
- Crew's photographs, videos and written notes of before and after the incident
- Evidence of the familiarisation of the bridge team with the equipment
- Maintenance records of bridge equipment
- VTIS recordings
- Bridge layout
- PSC inspections for previous 12 months
- ISM audits, internal and external
- Navigation audits.

Evidence specific to collisions

It is generally not possible to collect any evidence from the other vessel or vessels, if more than two are involved. Evidence from other vessels is normally disclosed at a later stage through the respective legal advisors.

In addition to the relevant general evidence, the focus should be on evidence that the lookout was maintained and the first sighting of the other vessels involved in the collision or near-miss. Evidence of bridge resource management will be required.

The surveyor should try to collect the following additional information from the vessel:
- Original working chart or ECDIS
- Chart correction log
- Radar being used and ranges
- Radar settings and performance (clutter, etc.)
- Tracking of targets on the ARPA and any manual plotting
- Lights observed and any changes in the aspect on the other vessel(s)
- AIS information of the vessel(s)
- Light and sound signals made, seen or heard
- Presence of other vessels in the vicinity of the action taken
- Proximity of any navigational danger and its effect on the ability of either vessel to take evasive action
- Any alterations of course and speed for navigation purposes during the passage, such as waypoints
- Any local navigational regulations applicable to the vessels, if the collision occurred in harbour or monitored coastal waters, channels or straits
- Any VHF communication with the other vessel, especially if there was any agreement on the action that was to be taken by one or both
- Any blind sectors on radar and any obstructions in the bridge view forward, such as crane parking, masts, deck cargo
- Any instructions, recommendations or advice provided by VTIS (if in VTIS area)
- Pilot's advice, if the collision occurred with a pilot on board.

Chapter 5
Guidelines for Collecting Maritime Evidence

Evidence specific to groundings

In addition to the relevant general evidence, the following should be obtained:

- Location of grounding and whether it was on a charted course or away from it
- Changes in position after the initial grounding
- Ballast and stability condition of the vessel at departure from the previous port and at the time of incident (if different)
- Draught survey reports at previous port (if applicable)
- Squat tables
- UKC used by the vessel in the preparation of the passage plan
- Edition of the chart (ECDIS or paper) in use and corrections applicable under *Notices to Mariners*
- Any history of previous groundings of other vessels in the area
- Tide tables and calculations
- Direction and rate of any current
- External and internal soundings taken after grounding
- Compass error
- If grounding occurred while manoeuvring in or out of a port, the port procedures for manoeuvring in the area and pilot's advice
- Pilot's experience, working hours and involvement in any previous incidents
- Turning area available, if grounding occurred while turning
- If tugs in use, their type, bollard pull and position at the time of the incident. Port requirements for number and bollard pull for size of vessel
- Any influence of overtaking or passing traffic
- Any time constraints that might have caused the pilot to hasten a manoeuvre or transit
- Weather thresholds for manoeuvring under port procedures
- Aids used for navigation
- Position fixing system used by the vessel and position fixing frequency
- Use of parallel indexing, if applicable
- Bathymetric charts of the area (held by the port authority if the grounding occurred in the vicinity of the port)
- Frequency, type and method of surveys undertaken
- Port AIS data.

Evidence specific to contact with fixed and floating objects (FFO)

Contact with FFOs often results in large claims. FFO damages occur within the limits of a port and offshore structures, mainly during berthing or unberthing manoeuvres. Such incidents often occur with a pilot on board.

Allegations of damage to berths, fendering and underwater structures can sometimes be received by a vessel after it has left the port. It is important to collect any evidence that may show some pre-existing damage.

Chapter 5
The surveyor's perspective

In addition to the general evidence listed earlier, evidence will be required to establish how the manoeuvre which resulted in the damage or alleged damage was carried out and how it compares to normal manoeuvres in that area. Evidence will also have to be collected ashore. In the absence of any engine or steering failure, the prime factors are likely to be the speed of approach to the berth or the angle of approach. Evidence of the type and number of tugs used, their power, whether fast, and their position, will also be required. The VDR will form the foundation for the reconstruction of the incident. AIS data can also be obtained in many cases and used for reconstruction. Historical AIS can also be used to ascertain what manoeuvres were used previously and the way transits were conducted.

VDR audio may be helpful to check the pilot's advice and the handling of tugs. As the Master and pilot are often standing at the extremity of the bridge wing during the final approaches to the berth, audio recording of their interaction and pilot's communications may not be recorded or be inaudible and must be supplemented by statements.

It is also important to check if any harbour or construction work (such as dredging) was being carried out in the area of the incident or berth. This could force a vessel to deviate from its normal approach plan.

If the allegation of damage has been made after the vessel has left the port, it will be necessary to visit the vessel at the earliest opportunity, preferably at the next port. It will be necessary to inspect the hull to check for any abrasion or contact damage in the area which is likely to have contacted the structure. If the suspected area is under water, then a hull survey is required, which the surveyor should witness and obtain the report and video recording.

Conditions during manoeuvring

The location of superstructures on a berth apron (cranes, loaders, chiksans) should be checked. For moveable equipment, such as gantry cranes, it will be necessary to collect evidence of their location along the berth during the manoeuvre. Port procedures for positioning of equipment during berthing and unberthing should be obtained.

If the port or berth has CCTV recordings, these should be obtained. In some cases, the port or terminal may not release the recordings initially but will allow the surveyor to view them. This opportunity should be taken and notes made on what has been observed.

It is also important to collect information on the location and mooring patterns of vessels in the immediate vicinity, as the space available may have a bearing on manoeuvres. Bollard locations are important factors. If a turning was undertaken, the surveyor must obtain evidence from the port of the turning areas permitted for the size of the vessel so that this can later be compared with the charted information available to the Master. Evidence should also be obtained about the type of fenders used, their number and location along the berth.

Chapter 5
Guidelines for Collecting Maritime Evidence

If a vessel breaks free (partly or fully) and causes damage to the berth or damage is caused due to ranging alongside in adverse weather, then evidence will be required of the existing conditions and the port's systems. The mooring spread deployed should be verified, together with the condition of the mooring ropes, method of securing and winches. Ship's drawings of mooring points, winch/bitts and lead capacities, certificates for mooring ropes or wires (including any tails and shackles), winch manuals, maintenance undertaken, method of use, settings required and used are required.

Damage to submarine pipelines and cables, which normally occurs at anchorages, will require plans of the location of the pipeline or cable. These should be obtained from the port or other relevant authorities, and verified against the information available to the Master from navigational charts. The vessel's positions at the anchorage and the observations when the anchor was heaved should be verified. Evidence should also be collected to ascertain whether the port instructed the vessel to anchor in a particular position. In cases where the vessel dragged anchor and fouled the underwater structure, evidence of anchor watch procedure and the action taken will be required. It will be necessary to obtain positional data of vessels in the vicinity at the time.

Wash damage

Such investigations require attendance on the vessel which is alleged to have gone past at a high speed or close proximity, as well as the vessel which is alleged to have been damaged. Such incidents usually occur within the confined waters of a port or river.

For a surveyor appointed on behalf of the ship which was passing, obtaining all the information or even gaining access to the vessel alleging damage may not be possible. The evidence is likely to only be disclosed as part of the process of litigation.

The main evidence to collect from the vessel accused of causing the damage would be the VDR data, if saved. Obtaining an AIS reconstruction in such a case would be very beneficial. If not, ensure that ECDIS, GPS, and echosounder data is collected.

The passage plan of the vessel and the information exchange with the pilot would be key evidence, in addition to the establishment of the vessel's actual course and speed data. It is not unusual for many berths, terminal and ports to have CCTV monitoring of operations.

The latest bathymetric charts of the berth area and the water beyond must be obtained from the port if it will release them. It will also be necessary to check the size and draught of vessels the port generally permits to moor at the berth and the port's speed limits in the area, if any. Evidence of other traffic passing or overtaking in the vicinity is important because this could have caused the passing vessel to go closer to the berthed vessel(s).

Evidence must also be obtained of any similar incidents off that berth or other areas of the port. It also needs to be verified whether vessels in the immediate vicinity of the damaged vessel were affected. That is particularly important along a long, straight quay wall, but also on dolphin-type terminals.

Chapter 5
The surveyor's perspective

While evidence collection on the damaged vessel may be limited, the exact position of the vessel before the incident should be obtained, and that along with the draughts will help in creating a 3D mooring model.

Apart from the speed and proximity of the passing vessel, the effectiveness of the mooring spread of the damaged vessel, condition of the mooring ropes (type, capacity and certification) and their tautness at the time of the incident are critical elements. Any evidence that shows whether a vessel of the size and type involved could have berthed safely at the berth and how the damage vessel was actually berthed would aid the investigation.

image: Danny Cornelissen (www.portpictures.nl)

THE NAUTICAL INSTITUTE

Chapter 6

The lawyer's point of view

By Michael Mallin and Jack Hatcher

In the event of a significant casualty, the involved parties will instruct lawyers because any such casualty will almost always result in litigation, often multiple litigation, as the various involved parties try to protect their position or recoup their losses. The outcome of the litigation will generally depend on the facts, and facts not captured at the time tend to be facts that are lost for ever. Lawyers are used because they should bring to the investigation an understanding of the legal issues that are likely to arise, and therefore what investigations might be required. Another reason for using a lawyer is that statements and other evidence gathered by a lawyer are normally given greater credibility by the Common Law courts, and the reason for this is that for a lawyer to knowingly or recklessly adduce misleading or false evidence to the court, or allow a witness to do so, or to write the witnesses' stories for them, amounts to serious professional misconduct. Lawyers are obligated to take greater care than others to ensure that the evidence they put forward is true and accurate. A lawyer who breaches this rule has been held by an English judge to be *"not fit to be a member of any part of the legal profession"*. Such lawyers are likely to have short careers.

Because the London Admiralty and Commercial Court remains the world's largest centre of choice for marine disputes, the lawyers will frequently be selected from one of the small group of firms of English solicitors that specialise in handling maritime casualties. Alternative jurisdictions of choice with similar (but not identical) 'common law' legal systems include the United States, Singapore, and Hong Kong – the last two are where the largest UK shipping solicitor firms have satellite offices.

The majority of these specialist firms use former mariners to collect casualty evidence, many of whom are also qualified solicitors. These mariner/lawyers generally bring valuable skills to the evidence collection process, including a combination of practical knowledge based upon service on a variety of ship types and the ability to appreciate what is legally significant.

In most significant cases, a mariner/lawyer will board a vessel immediately after a casualty to collect evidence. A separate but important task will be to assist the Master in dealing with the demands of the numerous parties who will also be trying to collect evidence, often to use against the shipowner. It is also common for representatives of the flag state, port, or coastal states to attend. It can be very difficult for the ship's staff to know with whom they should cooperate and how to deal with the multiple and simultaneous demands.

Chapter 6
Guidelines for Collecting Maritime Evidence

Each of the interested parties will be demanding a piece of the Master's time (often they each seem to want all of that time), during a period in which Masters will be both busy and extremely stressed, and often, also very tired. Regrettably, some may not be entirely frank about which interests they represent in order to try to elicit information from the ship's staff they might not otherwise have been able to obtain. The attending mariner/lawyer can advise the ship's staff and their owners, and protect and guide them through this intimidating and often confusing maze.

Traditionally, the mariner/lawyer's onboard investigation involved a series of in-depth interviews with the relevant ship's staff and a thorough analysis of the ship's documents. These included the bell book(s), logs, charts, and, where applicable, the engine room data printout records. The originals (or if not practicable, copies) of all relevant documents would be taken into safe custody. The lawyer/investigator would test whether the oral evidence of the witnesses was both credible and consistent. In cases involving navigation, this involved the use of parallel rulers, dividers, the ship's manoeuvring data, the paper course recorder trace (if any), and the results of the physical angle-of-blow survey to test whether the courses, speeds, bearings, and distances reported by the witnesses actually 'worked'. The reality was that the witness evidence rarely fitted either the factual scenario or the documentary evidence, and discussions would be held with the relevant witnesses to try to identify inaccuracies.

Nevertheless, it was common, even where the ship's witnesses had done their best to be truthful, that when the parties' lawyers eventually met to negotiate a settlement, the two sets of evidence were found to differ so substantially that it was difficult to see – for example in the case of a collision – how the two ships ever came into contact with each other; or in the case of a grounding, how the ship ever came into contact with land at all.

This traditional system, which had been in use for generations, has recently changed beyond all recognition – even since the publication of the third edition of *The Mariner's Role in Collecting Evidence* only 11 years ago. This change is particularly evident in navigation casualties such as groundings and collisions and is due to the widespread introduction of comprehensive electronic data recording systems in recent years. These systems have substantially transformed how the lawyer/investigator will collect and test the evidence.

Electronic evidence

Today, the sequence of significant events can often be pinpointed to the second. In many areas there is shore-based (or even satellite) AIS coverage, producing data which can be purchased from commercial organisations within hours of a collision occurring. The investigating mariner/lawyer will often have this evidence before even boarding the ship, enabling discussion of details with the witnesses before any electronic data from the ship is downloaded. VDRs now provide accurate contemporaneous evidence of the own ship's precise pre-casualty positions, courses, speeds, helm and engine movements, and the angle of blow. The VDR will have recorded screenshots of one or more radar and ECDIS displays, often in addition to the AIS evidence from other vessels involved, including their exact courses and speeds throughout.

No longer does it take several minutes, after those involved have dealt with the immediate emergency, to manually (and sometimes inaccurately due to the delay) record the time and position of a collision. Pre-grounding positions can no longer (sometimes!) be plotted on the chart after the event has occurred; nor does the evidence of the other ship's bearings and distances now depend on witness memory alone. Now, instead of simply interviewing the crew, recording their evidence and cross-checking it using the limited tools traditionally available, the mariner/lawyer collecting the evidence must also possess the skills to download and interpret the extensive contemporaneous electronic evidence available.

In addition to the VDR, automatically-generated evidence often available these days includes the ECDIS, GPS printouts, AIS data printouts, course recorder, echosounder recordings, tank soundings, and engine room alarm systems. The data retrieved from these electronic systems is most reliable as it comes from an independent, incontrovertible source with exact time stamps, which greatly assists the investigator in verifying the information provided by the ship's staff or obtained from various manually maintained records.

Disclosure of documents

Under the English and similar legal systems (including that of the United States), if a document is relevant to the case, it must be disclosed to the other parties in any resultant litigation, no matter how sensitive or commercially confidential. In most cases, only mariner/lawyers' own work product, such as witness statements prepared by them and written exchanges with them, or documents generated for the dominant purpose of litigation, will be privileged from such disclosure. Reports, notes, or internal office emails and memos prepared by the ship's staff or attending shore staff regarding the incident will have to be disclosed; as will any reports to third parties (such as an oil major or flag state).

Disclosure (sometimes called discovery) rules are not straightforward. For example, if the DPA attends the investigating lawyer's interviews, the lawyer's interview notes will be privileged but the DPA's will usually not be. The advice of the investigating mariner/lawyer is of significant benefit to shipowners wishing to avoid the very real possibility of accidently prejudicing their position during the evidence collection process.

Witness evidence after an emergency is notoriously unreliable. Accurately recalling events is extremely difficult. Furthermore, in the hours (or even days in remote areas) between the incident occurring and the arrival of the investigating mariner/lawyer on board, the witnesses will have repeatedly analysed the incident and reconciled in their minds what they believed occurred. They will undoubtedly have discussed the incident with their colleagues, and to be influenced by what other involved witnesses say is human nature.

Chapter 6
Guidelines for Collecting Maritime Evidence

For example, watchkeeping officers can, with the best of motives, agree on the distances and bearings of relevant objects or ships which they observed jointly. However, even with the best intentions, this may not be correct, and if, when the electronic records are downloaded, they are shown to be wrong, the reliability of these witnesses may be fatally compromised if they have reduced those incorrect assertions to writing (and therefore disclosable evidence) before reviewing available electronic data. They may be accused of collusion and will often be discredited for cross-examination at trial.

Therefore, it is now prudent for the investigating mariner/lawyer to download and analyse the readily available electronic evidence, so that it can be discussed with the witnesses during their interviews. Otherwise, once electronic evidence does become available, it will usually show inconsistencies in the witnesses' recollections and follow-up interviews may be required, resulting in inevitable delay and additional expense.

Therefore, today's onboard investigating mariners/lawyers need to have the skills to access, interpret, and save the ship's electronic data. They must maintain and develop these skills through regular voluntary training courses in modern radar and ECDIS. Unfortunately, the various manufacturers of marine equipment do not all use the same software. Quite often the software will first have to be obtained or downloaded. The onboard assistance of a manufacturer's technician is invariably still required, particularly in relation to the VDR downloads.

It is also important to preserve the integrity and the 'chain of possession' of the electronic evidence downloaded and the evidence of the lawyer as to the origin of the data will normally be accepted by the court in circumstances where the evidence of one of the interested parties may not be.

Analysing electronic data

The audio VDR record will inevitably contain terms which would be incomprehensible to a non-mariner. Therefore, when a complete and accurate transcript is required for legal purposes, the audio record is best transcribed by someone who has previously served at sea. For example, our firm now employs an IT-skilled paralegal who was formerly a junior ship's officer as an Admiralty Technical Assistant and this may become common. She works in conjunction with the firm's investigating mariner/lawyers worldwide to assist in the downloading, handling, and analysis of the electronic data. By regularly attending external manufacturers' courses, she remains up to date with the latest developments and software in the maritime electronics field. During the evidence collection process technical office-based back-up is therefore available to the onboard mariner/lawyer.

If shore-based AIS data is available, computer-generated plots, scale diagrams and pictorial simulation can be swiftly produced for use during the witness interviews where necessary. Together, the investigating mariner/lawyer and the witness will listen to the audio recordings taken from the bridge microphones and the VDR recording of VHF traffic. This can be time-consuming because audio evidence usually requires several replays before

Chapter 6
The lawyer's point of view

it can be understood, particularly when there are several witnesses and often different languages involved. Without this process, however, it is not possible to obtain complete and reliable evidence, and the witnesses' statement evidence will be inaccurate.

We also analyse the electronic evidence downloaded by the mariner/lawyer from the ship's systems, presenting the results in diagrammatic or pictorial form to both the shipowner and the insurers, so that they too can better understand how and why the incident occurred.

As a result of technological data retention, the onboard mariner/lawyer investigator often focuses less on what happened (regarding the ship's physical movements and manoeuvres) but rather, why it happened. Furthermore, the investigator will explore the thought processes of the Master and relevant officers and other witnesses leading up to the incident. Lines of enquiry might include why they waited until a specific time before taking action or why the ship's position was not fixed earlier. The mariner/lawyer may try to ascertain why, given the circumstances, the crew ventilated (or did not ventilate) the cargo or why they selected a particular radar or chart scale, the course track, or way-point or setting on the ECDIS. There may be a good reason which is not otherwise apparent.

They may ask for details of the ship's onboard fire drills. These are all pieces of evidence which will not only affect the determination of who is responsible for the incident, but also the attitude of the court; and this is evidence that cannot be obtained solely from the vessel's electronic records. The time required for the proper onboard collection and review of evidence by the mariner/lawyer is now probably twice as long as it used to be, but the much greater certainty and independence of that evidence now means that far more disputes settle at an early stage and without litigation, and therefore far more economically overall.

Machinery damage and fire

The position can be different when the mariner/lawyer investigator is dealing with machinery damage and fire cases. These often lead to insurance, cargo, and other claims. In catastrophic machinery damage cases the machinery data recordings will often not indicate the cause of the damage and the ship's engineers will have to be interviewed about the event and about maintenance procedures. The investigating mariner/lawyer will often be accompanied by a consultant marine engineer who will conduct a forensic examination of the damage and assist with the crew interviews. Laboratory analysis of engine parts may sometimes be necessary.

In fire cases, there may be minimal relevant evidence surviving in the ship's automatically generated records. The investigating mariner/lawyer will work with a forensic fire scientist, and in these cases the crew witness evidence often remains of traditional importance. This is particularly true for insurers' lawyers in the relatively rare cases where insurance fraud is suspected. Fire is a particularly suitable device for fraudulent insurance claims, because, for historical reasons, the insured only has to prove loss by fire to succeed in his claim; the insured does not have to prove how the fire started or spread. In such cases the investigating mariner/lawyer will need to compile a full matrix of the activities of the whole crew by interviewing everybody who was on board at the time.

Seaworthiness

In a typical collision case, the collection of evidence by the investigating mariner/lawyer may well be limited to the actions of those on board. However, many other casualties require the collection of evidence from other sources, including from the ship owner or manager. Most marine insurance policies and P&I Club rules include 'warranties', which are promises by the insured (also called the assured). A breach of warranty can prejudice the insured's insurance cover. Also, under English insurance law, if a ship puts to sea in an unseaworthy condition with the "privity of the assured", the underwriter will not be liable for any casualty as a consequence of that unseaworthiness.

In most contracts for the carriage of cargo or charterparties, the shipowner will be liable for the cargo and/or charterers' losses resulting from the vessel's unseaworthiness "before or at the beginning of the voyage" unless the shipowner can prove that he (and all his servants, agents and contractors) has exercised "due diligence" to make the ship seaworthy. Due diligence failure is equivalent to negligence. Therefore, a causative error made by anyone in the shipowner's office ashore or a third party repairer employed by him may expose the shipowner to liability. Generally, the shipowner can limit liability to third party claimants to an amount which depends on the gross tonnage of the ship. However, that protection can be lost if the conduct of senior shore staff is especially reprehensible – being at least reckless, knowing that this would probably cause the loss.

If the amount of money at issue or the nature of any disputes justify such investment of time, the evidence relevant to these issues will have to be collected. This may, for example, require collecting evidence regarding the behaviour and procedures of the shipowner's office ashore as well as the actions of those concerned with the management and operation of the ship. The nature of the casualty will of course govern the extent of the evidence required. In most casualties, the shipowner will wish to demonstrate the competence of the ship's staff. This will typically require the collection of evidence relating to their recruitment, training, and the supervision of onboard procedures by the office. A review of the shipowner's SMS will be required, and evidence will need to be collected to show that these procedures have been applied on board, as well as how they are supervised and audited. Most mariner/lawyer investigators have ISM audit training.

Chapter 6
The lawyer's point of view

Shore-side systems and records

Today, office records tend to be stored electronically (and almost indestructibly) and are often more detailed than previously. Data is transmitted automatically from the ship to the office. All this is evidence, and it will be reviewed and dealt with by the investigating mariner/lawyer as part of the evidence collection exercise. Additionally, follow-up interviews of shore staff may be conducted.

Where the casualty is the result of machinery failure, a detailed investigation of the shipowner's maintenance records, procedures, and ISM Code Section 9 audits will be necessary. The investigation can be complex. It often includes a review of running hours, maintenance reports, defect reports from the ship, internal and external audits, reports of the attending superintendent, internal management memos, the correspondence between the ship and the shore, and the company's policies and practices confirming the supply of spare parts and external technical assistance in relation to maintenance of the ship. In the event of litigation, all of these documents may have to be produced to opposing parties in due course as part of the disclosure process (see below), and any problems revealed will have to be taken into account during the evidence collection exercise.

Additionally, a review of the shipowner's responses to incidents of a similar nature across the entire owned or managed fleet will sometimes be necessary. The mariner/lawyer collecting the evidence will then review this information and the relevant personnel may need to be interviewed, all to prove that the shipowner has complied with the obligations stipulated in the insurance contract or the contract with the cargo or charterers.

The disclosure process, which requires each party to produce to the other all relevant documents that are within their possession or control, including internal and/ or confidential documents whether generated before the casualty or at any time afterwards up to the time of the trial, now includes all evidence stored electronically. As part of their duty to the court, the parties' respective solicitors are required to conduct a reasonable search for relevant documents. This usually involves an examination of current data, including data stored electronically. Most common law systems only require disclosure which is 'proportional' to the issues and the nature of the case, so that in a smaller case, the search for relevant documents which a party is obliged to do will be less extensive.

In appropriate cases, however, the court may order a forensic collection of a party's electronic data by specialist IT consultants under a lawyer's supervision. This may include the imaging of electronic records and metadata, and will comprise not only current documents, but also previous drafts and deleted documents, with searching often by means of keywords.

The burden upon shipowners, and therefore lawyers collecting evidence on their behalf in order to defend their interests, is often particularly heavy in cargo claim cases. This is due to the obligation to prove that due diligence has been exercised by the

Chapter 6
Guidelines for Collecting Maritime Evidence

shipowner. Essentially, the owner has to prove the negative – that there was no causative negligence on the part of his servants, agents, or sub-contractors relating to the casualty.

One of the authors started working as a mariner collecting evidence over 30 years ago. The collection of evidence by the mariner/lawyer today is in some respects a whole new world. Traditional skills are still required, however. Mariner/lawyers must be skilled investigators and interviewers and be sensitive to the position and cultural values of the people they are dealing with. However, they must now also possess the skill set to cope with the whole range of electronic data generated by the modern ship and its owners.

Chapter 7

A marine claims broker's perspective

By David Keyes and Ivor Goveas

Marine claims brokers act solely on behalf of the shipowner and always have the shipowner's interest at the forefront when dealing with insurers. They can be advocates for the shipowner, representing the shipowner in all communications with the insurers. The claims broker's role has become more diverse in recent years with intervention in casualties across the different elements of marine insurance: hull and machinery (H&M); loss of earnings; loss of hire insurance; and P&I. When a casualty occurs, the marine claim broker's key function is to coordinate the many different experts instructed by the parties and to minimise any actual or perceived conflicts of interests between the parties or their experts.

In most insurance regimes, the onus of proof remains with the shipowner to prove the cause of the loss being an insured loss and to fully demonstrate and evidence to the insurer's satisfaction the reasonable extent of repairs and the reasonable cost of the repairs. As the agent of the shipowner, the marine claims broker plays a significant role in ensuring that the documentation secured at the time of the maritime casualty is of sufficient quality to ensure that the shipowner's onus of proof is satisfied.

A broker's role is to facilitate the indemnity claim under the insurance policy and, where conflicts or disputes arise, to bring shipowner and insurer together to find an acceptable solution. However, as in many other areas of contract law, a solution may not be readily available and disputes may end up in litigation. Therefore, the quality and reliability of evidence gathered for insurance purposes should ultimately be sufficiently robust to face the test of a court of law.

Evidence gathering at the earliest stage of a casualty is essential to ensure proper understanding and correct identification of causation and determination of liability.

The marine claims broker is often the first point of contact following a casualty, so that all involved insurers can be properly notified.

Several factors dictate the level of cooperation that may be possible between the various insurers involved in an incident. This in turn would also dictate the level of confidentiality required to be maintained between the various insurers. The marine claims broker is in the ideal position to facilitate this.

The Master's first task is to attend to the consequences of the casualty, including care of injured personnel, and control of damage to the ship and the environment. It is also necessary to notify insurers, who will be more concerned with rectification of the consequences of the incident, which may include:

Chapter 7
Guidelines for Collecting Maritime Evidence

- Pollution (P&I)
- Damage to cargo (P&I)
- Damage to insured property (H&M)
- Damage to other property (P&I and H&M).

In addition, in order to determine the extent of their liability, insurers may require the instruction of surveyors, experts and lawyers in order to collect evidence and provide advice as to causation, liability and quantum.

Some insurers may require different types of evidence, subject to the risks on cover, but there will invariably also be evidence that is common to all policies. If collection of evidence was undertaken by insurers in direct contact with the insured, it is likely to result in duplication of efforts by the insured. The broker can coordinate the collection and proper distribution of evidence from the insured to the various interested insurers.

Acting as a coordinator also puts the broker in a position to raise queries with insurers about evidence requested where its necessity may not be readily apparent. Such correspondence with the insurer can then be passed on to the insured along with the request for evidence, which may help in a smoother process. It is also likely that some of the information requested is already in the possession of the broker.

Average adjustment

The marine claims broker will invariably be involved with the drawing up of an average adjustment, either undertaking the task in-house by drawing up a statement of claim or liaising closely with an average adjusting firm. This will include providing the average adjuster with the information required for proper and fair adjustment of the claim. The broker being involved with the collection and dissemination of evidence from the very beginning of the claim process would greatly assist in the efficient execution of the adjustment process.

There is some common information and documentation that will be required for all casualties that may develop into an insurance claim. This includes, but is not limited to:-

- Copies of log extracts and reports, which should be made ready for examination by the insurer's surveyor
- Any damage to cargo should be noted, giving the extent and cause of damage and identifying clearly the cargo in question
- Consumption of ship's fuel and stores should be noted, eg on tank cleaning, repairs or deviation to repair port
- Claims by third parties for damage caused by the ship or to cargo should be referred promptly to the local P&I representative, and notified to the shipowner
- Past log book entries leading to the incident
- Maintenance records
- Master's and chief engineer's statements
- Other witness statements

- Classification society surveyor's report
- Superintendent's report
- Class maintained certificate issued by the relevant classification society confirming the vessel has been maintained in class from inception of the policy to the date of the casualty
- ISM Document of Compliance (DOC) and Safety Management Certificate (SMC)
- H&M insurers' surveyor's advices and reports which may be available to the marine claims broker.

Specific documentation and evidence should be secured for the following types of casualty.

Machinery breakdown

This category of claim tends to generate most insurance disputes as they are often caused either by hidden defects in the machinery, latent defects or crew negligence, and are open to opinion by the opposing parties. It is therefore critical that the evidence that is collected following a mechanical breakdown is accurate, reliable and detailed.

Damaged parts should be retained and preserved and maintenance records and running hours should be made available; appropriate samples (eg lube oil, bunkers) should also be taken for analysis, if necessary.

Case study 1

A vessel had a catastrophic failure of an auxiliary engine when a piston shaft broke. The surveyor's initial investigation indicated that the chief engineer had repeatedly pressed the 'start' control and could have started the engine up to seven times with additional damage occurring each time. Due to ambiguous wording in the excess/deductible clause, the insurers were able to argue that an excess/deductible could be applied each time the chief engineer pressed the start control, leaving the possibility that seven excesses/deductibles could be applied to the claim. The shipowner would not receive an insurance indemnity in respect of the casualty.

Brokers studied the engine log books and discovered that the chief engineer had initiated the start sequence several times but the auxiliary engine had started only once. This demonstrated that all the damage occurred in one action of the chief engineer and one catastrophic failure of the engine, allowing the shipowner to successfully make the claim.

General average

A claim that gives rise to general average may have both H&M and P&I implications. The average adjuster will require copies of the charterparty and bills of lading to determine the regime applicable to the GA adjustment. This will also be required to identify charterers and cargo interests for the purpose of collecting GA security.

P&I insurers will be interested in making an early assessment of evidence that they may require to defend any claim of unseaworthiness by charterers and cargo interests in an attempt to avoid paying their GA contribution or put themselves in a position whereby they may be able to negotiate a discounted contribution.

Lawyers and surveyors may also be instructed on behalf of both H&M and P&I insurers and will also require access to certain evidence. The insured will require an overview of all activity and will also be involved in providing information as required by the various parties.

Much of the information required will be available on board the vessel, and requests may go directly to the Master. In addition, there will also be requests for information from opponents. It is, without doubt, extremely important that such requests are properly considered before being complied with.

Given the number of parties involved in such an exercise, it would be sensible to have a focal point through which information flows in all directions. This is information that will be used as evidence by various parties including surveyors, lawyers and adjusters. Additionally, it will be used by insurers to determine that there are no issues with cover in respect of the relevant policies, and by charterers and cargo interests to determine whether a breach of contract by shipowners was causative of the incident.

The documents required in cases of general average vary considerably according to the nature of the casualty. The following cover the majority of cases.

Vessel resorting to a port of refuge

- Log extracts and reports from the Master or other parties showing the dates and times when the vessel deviated, arrived at port of refuge, left port of refuge and regained its position to resume the interrupted voyage
- Any survey reports, whether held on behalf of insurers, owners, the classification society, or in the general interest, dealing with the vessel's resort to the port of refuge and any repairs carried out there
- Details of any repairs carried out at the port of refuge, stating whether they were temporary or permanent, and how much of the repair account represents the excess costs of overtime worked by repairers
- Details of any shifting or discharge of cargo at the port of refuge, stating whether such shifting or discharge was necessary in order to allow repairs necessary for the safe prosecution of the voyage, or for the common safety, or for re-stowage
- Superintendent's report if attending at the port of refuge
- Details of fuel and stores consumed in deviating to the port of refuge, while detained there, and in regaining position, together with details of the cost of their replacement (or relevant off-hire statement)
- Substituted expenses, such as costs for forwarding cargo to destination on another vessel, to the extent of costs that would have otherwise been incurred were it necessary to store the cargo ashore during repairs and re-loaded on the vessel after repairs.

Fire on a vessel
- Survey report showing division of damage between fire and efforts to extinguish the fire. The same division should also be made in surveys of any similar damage to cargo
- Details of any fire-fighting depletion, such as refilling extinguishers, CO_2 bottles.

Groundings
- Survey report dividing the damage found between that caused by grounding and that caused by refloating
- If the vessel has been refloated with tugs, details of the salvage award and relevant legal costs, or if the salvage services have been rendered under contract, a copy of the salvage contract and the relevant accounts
- Any lightening attempts
- Salvor's activities
- Equipment and personnel employed by salvor
- Status of vessel during the refloating operation, eg soundings, extent of damage
- Assistance given by crew and equipment used during the refloating
- Weather conditions during and subsequent to salvage operations, and any other influencing factors
- Actual or potential escape of cargo or bunkers.

Cargo
- Manifest of the cargo on board at the time of the accident
- Copy of the bills of lading showing the front and reverse sides
- Details of the out-turn of cargo delivered
- Reports of survey on the cargo held directly following the casualty or at the port(s) of destination.

Collision

In order to be effective, this type of claim requires the closest liaison between H&M and P&I. H&M is primarily concerned with the damage to the hull and machinery and also (under English insurance conditions) three-quarters of the liability to the third party vessel. P&I is primarily concerned in a quarter of the liability to the third party vessel and potential ancillary liability issues such as injury to personnel, pollution, removal of wreck and cargo liabilities. Brokers can liaise with experts, including surveyors and lawyers, ensuring there is no duplication of instructions and that clear lines of communication are established.

In the event of a serious collision, it is likely that a lawyer will be appointed immediately to protect all of the ship's interests. See Chapter 6 for further details.

Chapter 7
Guidelines for Collecting Maritime Evidence

Case study 2

Cargo was transhipped and carried to destination on a substitute vessel to facilitate repairs to the original carrying vessel following on from a casualty. This was undertaken as a substitute to storing the cargo off the damaged vessel during repairs. The substitute vessel did not have cargo heating capability. When the cargo arrived at port it was at the incorrect temperature for discharge, causing a delay in berthing and incurring demurrage in an amount of approximately $500,000. This amount was admissible in GA, but the adjuster had not been made aware of this cost as the assured had not realised that this was claimable. This came to light in follow-up discussions with the broker, and the amount was subsequently included in the adjustment.

Casualty-related information, documentation or evidence may only be required several years after the casualty. Therefore, an owner should take particular care to ensure that the evidence, once collected, is safely stored in both its original paper form and electronically, and carefully referenced. This will ensure that when the evidence is required it can easily be retrieved without major effort or re-work.

Chapter 8

Evidence for insurance claims

By Peter Young

Most insurance claims concern cover for the liabilities of the shipowner, loss of earnings or damage to the vessel and its associated machinery.

These insurances, including P&I Clubs, are governed by English Law with some exceptions. The other significant applicable laws are Norway under the Nordic Insurance Plan, US Federal Law under the American Institute Clauses, Germany under the DTV clauses but also Japanese, Korean and increasingly Chinese law that generally follow English legal principles.

Insurance terms and conditions

All English insurance policies contain the provisions of the Marine Insurance Act 1906 (MIA1906). A new Insurance Act 2015 came into force in 2016 that removes certain sections of the MIA 1906. However, this mainly concerns amended requirements for disclosure of information to insurers when insuring the risk and some modifications to the effects of a breach of express warranties. P&I Clubs have opted out of the provisions of this Act, whereas the commercial market has generally accepted it.

Both H&M and P&I insurances accept the principle of 'proximate cause' as defined under S55 of the MIA 1906 in order to determine whether a claim can be made. The proximate cause is considered to be the dominant cause rather than that necessarily closest in time. This is distinct from the root, direct or contributory causes of insurance claims.

The collection of evidence is an important part of the process of determining the dominant cause. Sometimes, the dominant cause is not clear and experts from many disciplines are required to establish it.

Insurers universally accept that the laws of different countries may change or affect the liabilities the mariner would have under English Law. The IMO has codified the mariner's liabilities in certain areas through international conventions on pollution, carriage of hazardous and noxious substances, humanitarian protection for the seagoing mariner, and the prevention of accidents at sea.

P&I insurance claims

P&I Clubs insure liabilities relating to pollution, cargo (passenger, general, liquid and dry bulk), personal injury, stowaways, deviation to save life at sea, and damage to fixed and

Chapter 8
Guidelines for Collecting Maritime Evidence

floating objects, subject to certain financial limits. Separate insurance must be arranged to cover wreck removal and other liabilities in the event of war, sabotage, or terrorism.

Pollution

This could take the form of oil cargo or bunkers, or ship's waste (sewage, garbage, grey water, ballast water, engine exhaust discharge, slops, oily water and sludge).

A pollution incident can be identified in circumstances including:
- An oil sheen close to the vessel or originating from another vessel
- A trail of oil in the wake of a vessel
- Floating debris, odours or discolouration of the sea or dock water
- An unaccountable discrepancy in the oil record book
- In the US, evidence from a whistle blower
- Discovery of a potential or actual 'magic pipe' on board
- An accident, collision, grounding, stranding or sinking.

For a detailed list of suggested evidence, see pages 48-52 of *The Mariner's Role in Collecting Evidence – Handbook* (hereafter referred to as the handbook).

It is important to always maintain meticulous up to date records on board as false allegations have been made by whistle blowers from improper motives. Do not try to cover up an incident by altering or fabricating records retrospectively. This can lead to criminal sanctions against officers and crew, which may include being forbidden to leave the country concerned even before conviction.

A local P&I surveyor should be appointed as soon as possible after a pollution incident to investigate the circumstances. For a major incident, specialist surveyors, lawyers and oil experts from ITOP will be required.

It is now possible to identify the type and origin of oil in order to counter any false allegation against a vessel and also in circumstances where the crew are unjustly accused of causing pollution (as in the case of the *Hebei Spirit*).

Cargo

For a detailed list of suggested evidence, see pages 29-32 of the handbook.

Passengers

Passengers are regarded as cargo by insurers and are subject to the terms and conditions of their passenger ticket and the Athens Convention, which has increased limits on liabilities for passengers by virtue of the PLR protocol effective for European Union countries from 31 December 2012. The major exception is for US passengers where the liability regime is much higher.

Special attention should be given to collecting evidence of passenger incidents and a detailed investigation of the circumstances should be carried out, as they can result in

Chapter 8
Evidence for insurance claims

very high liabilities (as in the case of the *Costa Concordia*). The public nature of passenger casualties will also give rise to substantial media attention.

Incidents can often be caught by passengers and crew on mobile phones, iPads or cameras, so it is often worthwhile establishing what is available. Care should be taken when seeking to obtain copies from passengers as there may be data protection issues.

In the event of a major casualty, class, specialist surveyors, lawyers, medical staff and experts will be required to immediately attend the vessel (as in the case of the *Herald of Free Enterprise*). Their attendance will help minimise losses after the accident.

Liquid cargo

Most liquid cargoes have independent surveys upon loading and discharge which are a great help in avoiding major problems. The results of these independent surveys should always be checked. If not satisfied, carefully assemble the evidence in support. If there is no agreement then an alternative, preferably specialist, surveyor should be appointed.

In the event of an incident when there has been contamination, close examination is needed of evidence of the tank and pipeline cleanliness prior to loading, any transfers during loading, and of valves opened or closed during loading and discharging operations.

If cargo heating is used, obtain details and evidence of the condition of the tank cargo heating pipelines. For chemical and product tankers where the paint tank coating fails, details of previous cargoes carried should be carefully recorded.

For a detailed list of suggested evidence, see pages 36-38 of the handbook.

Dry cargo

Insurance claims, which can be very serious, arise mainly from:
- Improper loading plan causing severe hogging or sagging
- Loading rates excessive for the vessel's scantlings
- Cargo liquefaction causing instability
- Heating or spontaneous combustion of coal or grain
- Grain infestation due to poor fumigation and hold cleanliness
- Sulphurous corrosion caused by coal
- Poor loading or securing plan for heavy cargoes insufficient for heavy weather
- Poor condition of the hatch covers and seals
- Failure to secure all watertight doors to the vehicle decks
- Lack of knowledge of cargo carried in containers – incorrect manifest declaration
- Cargo shortage and damage due to pilferage, improper discharges or inaccurate tally
- Severe chemical reaction in fertilisers
- Cargoes that react with water, such as cement
- Tainted cargoes – improper loading plan
- Reefer cargoes: failed equipment (eg generators) or incorrect set temperatures
- Condensation and wet damage – container and hold ventilation

Chapter 8
Guidelines for Collecting Maritime Evidence

- Defective loading or cargo securing equipment
- Stevedore damage
- Incorrect voyage orders.

For a detailed list of suggested evidence, see pages 33-35 and 39-40 of the handbook.

Crew personal injury, fatalities and illness

Insurance claims mainly arise from:
- Entering a space not properly ventilated
- Slips and falls – carelessness, poor lighting and deck condition
- Inadequate heavy weather arrangements
- Equipment and structural failures
- Improper repair and maintenance procedures
- Drug and alcohol incidents
- Inadequate pre-employment medical examination
- Latent illness
- Industrial disease, such as hearing impairment and exposure to asbestos
- Excessive working hours
- Improper clothing
- Inadequate training and lack of knowledge
- Incorrect medication
- Third party liabilities (eg passengers, pilots, surveyors and stevedores).

For a detailed list of suggested evidence, see pages 19-21 of the handbook.

Stowaways

Insurance claims mainly arise from:
- Inadequate inspections prior to departure
- Lack of knowledge of high risk areas
- Inability to identify the country of origin of the stowaway
- Drug smugglers concealed on board
- Difficulties in returning stowaways to their country of origin.

Specialist advisers can be appointed to assist in resolving these issues. Early collection of the suggested evidence can be of considerable value.

For a detailed list of suggested evidence, see pages 23-24 of the handbook and The Nautical Institute publication *Stowaways by Sea*.

Deviation to save life at sea

This is a requirement of international law and P&I Clubs insure these costs.

For a detailed list of suggested evidence, see page 25 of the handbook.

Damage to fixed and floating objects

The local P&I correspondent should be notified and a local P&I surveyor will be appointed to inspect the damage to the fixed and floating object.

The collection of evidence is considerably assisted by port VTS and the vessel's VDR evidence.

Insurance claims mainly arise from:
- Blackouts on board due to a generator tripping out
- Pilot error – report required
- Negligent tug assistance
- Poor communication and data exchange with the pilot
- Pilot's lack of understanding of the vessel's manoeuvring characteristics
- Ranging damage – lack of preparation for weather, port characteristics and swell conditions
- Poor state of mooring ropes and arrangements
- Wash damage – excessive speed, insufficient knowledge of the wake characteristics of vessel.

For a detailed list of suggested evidence, see pages 52-58 of the handbook and pages 34-36 of Chapter 5.

Hull and machinery (H&M) claims

Most, if not all, major accidents at sea give rise to H&M claims, particularly collision, sinking, grounding, fire and all types of machinery failure.

Generally speaking, H&M claims are the largest in terms of frequency and severity, with the exception of major P&I claims that concern pollution, collision liability if not insured under H&M, wreck removal and passenger compensation. Owing to the efficiency of sea transport, notably containerisation, cargo claims have reduced significantly over the last 10-15 years.

In terms of H&M claims, navigation accidents (with the possible exception of fire and explosion) give rise to the largest claims and machinery failures are the most frequent. Machinery failures can lead to major navigation accidents.

A major accident will probably concern both H&M and P&I insurers. After notification of a major accident, many surveyors representing both insurers will be appointed to attend the vessel. In the case of collision there will be without prejudice (WP) surveys carried out by the opponent vessel's interests, plus representatives from class, the salvor, flag state and any coastal state affected.

Each of them will be seeking to collect evidence in their own interest. Consequently, it is very important for a lawyer appointed by the owner to attend and assist those on board at the earliest possible opportunity.

Collection of evidence should be carried out by those on board as early as possible, provided this does not interfere with any post-accident minimisation of loss, such as salvage.

Chapter 8
Guidelines for Collecting Maritime Evidence

Technology in the form of VDR, VTS, AIS (now available from third-party companies), alarm, bridge and engine room data loggers, greatly assists the collection of evidence.

For a detailed list of suggested evidence, see pp 52-60 and 62-64 of the handbook.

H&M insurance claims mainly arise from:

Collision
- Failing to take early positive action
- One vessel altering course to port
- Failing to keep a proper lookout
- Failing to appreciate a developing close quarters situation
- Pilot error
- Poor design (eg bridge window visual distortion).

Grounding
- Failure to keep a proper lookout
- Failure to closely monitor the vessel's position at sea and at anchor
- Machinery failure, such as blackout
- Incorrect charts
- Pilot error.

Fire and explosion
- Failure to recognise a hazardous situation
- Machinery failure, such as burst fuel pipe or crankcase explosion
- Inadequate hot work procedures
- Hazardous cargoes and spontaneous combustion.

Machinery failure
- Poor lubrication
- Defective parts
- Poor design
- Improper overhaul procedures
- Poor maintenance
- Repairer's negligence.

In addition to the checklists on pp 62-64 of the handbook, the following documents and support that will usually concern an Average Adjuster for an H&M claim may be required:

Deck and engine room (ER) log book:
- Covering the voyage on which the accident occurred from the commencement of loading until completion of discharge
- Covering the full period under repairs
- If the vessel deviated for repairs, covering the deviation passage to the repair port, and to the next port of loading after repairs

Chapter 8
Evidence for insurance claims

- Sea protest or ship's declaration, if made, including the cost
- Master's/chief officer's/chief engineer's casualty or damage report.

Reports or surveys of the following at each port where a survey is held or repairs effected:
- Insurer's appointed surveyor
- Class surveyor
- Owner's superintendent, if required
- Diver, including video, ultrasound reports if required
- Invoices for fees and expenses of the surveyors above
- Specifications prepared and tenders taken for repairs.

Shipyard invoices for:
- All damage repairs, both temporary and permanent
- Drydocking and general repair expenses
- Tank cleaning and slop disposal, if necessary
- Gas-freeing chemist and costs
- All owners' repairs conducted concurrently with damage repairs, including any cleaning and painting of the vessel's bottom.

Agent's general accounts together with all supporting vouchers:
- All ports where repairs were undertaken
- High pressure cleaning of tanks and oil cargo or bunker removal costs, if relevant
- If insured, details of monthly wages and maintenance of the crew, including overtime payments directly in relation to the repairs. There may be also hold cleaning in preparation for repairs, for example, when proceeding to the repair port.

If not included in the agent's general account, or if there is no agent, invoices covering the following at repair ports:
- Transporting the ship to and from drydock and/or repair berth eg pilotage, towage, boatage and riggers
- Compass adjusting on completion of repairs, if required.

Supporting details for special payments made to the crew:
- In connection with damage repairs
- Tank or hold cleaning or other work.

Master's statement of the quantity of bunkers used (if the daily consumptions are not shown in the ER log book):
- At ports of repair, during drydocking, undocking and repairs
- During the deviation of the vessel to the repair port and to the next port of loading after repairs (the deviation triangle). The ship might proceed to the discharge port after repairs and might, before arrival, reach an equivalent point to continue its voyage

Chapter 8
Guidelines for Collecting Maritime Evidence

- During periods at sea for tank cleaning and/or gas freeing
- During periods of sea trials after repairs
- Details of costs for bunkers taken at the next subsequent bunker port after repairs, or if the ship is on timecharter, an offhire statement evidencing the cost of bunkers deducted for the bunkers so used.

Details of costs incurred for replacing any stores, equipment, or spare parts:

- Lost, destroyed or damaged beyond repair
- Used in connection with repairs
- Used for tank cleaning
- When spare parts or equipment are specifically ordered due to the damage, costs covering their transportation to the ship, insurance and delivery at the repair port
- Credit notes for the scrap value, or proceeds of sale, of old propellers, tail shafts, or other parts condemned due to the damage
- Details of the hire of temporary generators, including additional equipment used and special fuel used, and additional insurance costs for hired machinery.

In collision cases (in addition to pages 54-55 of the handbook):

- Lawyer's advices of the recovery from the opponent vessel
- Lawyer's advices on any amount paid in respect of liability to the opponent vessel or other third parties, for example losses due to pollution
- SCR representative costs and any other pollution related costs
- Invoices for legal costs incurred
- Invoice for the bail fees and other costs for providing security or guarantees given to opponent vessel
- Invoice for insurance for providing security
- Invoices for the costs of WP damage surveys on opponent vessel
- Details of additional communications costs and other small expenses in connection with the accident or repairs
- Invoices for fees and expenses of shipowner's agents and managers
- Dates of payment and currency conversions of all costs
- H&M insurance terms and conditions, including deductibles
- Document of Compliance (DOC) and Safety Management Certificate (SMC)
- Class maintained certificate from commencement of policy period until date of accident
- Letter from the DPA confirming that the DOC and SMC are valid
- If required, evidence that the assured was in compliance with any related warranty in the policy (eg trading warranties in respect of ice damage)
- Details of any extra costs incurred in order to save delays.

Chapter 8
Evidence for insurance claims

General Average (GA)

Documents that may be needed in addition to those listed on pages 49-52 of Chapter 7 and page 60 of the handbook include:

Agent's general account, with supporting costs for all ports of refuge:
- Master's portage bill for the voyage
- If not shown on portage bill, the wages of the Master and any radio communication costs
- A note of the daily maintenance cost of the crew.

Any special vessel overtime sheets and details of work done:
- During the periods of peril (while aground, fire-fighting, etc)
- In carrying out emergency repairs
- Bearing up for ports of refuge and regaining position.

Details of bunkers used:
- During re-floating, pumping and fire-fighting etc
- Proceeding to the port of refuge and regaining position
- While at the port of refuge
- Details of the cost of bunkers taken following repairs. If on timecharter, offhire statements showing the costs of bunkers deducted.

In salvage cases:
- Details of the settlement of the salvor's remuneration or award, usually under Lloyd's Open Form (LOF)
- Amounts paid to salvors
- Details of legal costs incurred
- The Committee of Lloyd's and Salvage Arbitrators' costs, if applicable
- Salvor's legal costs
- Invoice for bail fee given as a guarantee to salvors
- Invoices for costs for insurance of the salvage guarantee
- Details of the insurance and invoice for providing the salvage guarantee
- Details of any special insurance for cargo discharged temporarily in order that repairs could be carried out. Records of what has been discharged.

If a surveyor is appointed in the GA interest, whether for ship or cargo:
- Copy of the survey report
- Costs for fees and expenses of the GA surveyor(s).

GA security documents:
- Average bonds
- Underwriters' or P&I Club's guarantee or letter of undertaking
- If deposits are taken, a list of depositors, with deposit receipts and information of which bank account these are lodged in. This could be huge for large container ships

- Copies of commercial invoices for the value of the cargo
- Particulars of the value of the property on board belonging to parties other than the owner and those concerned in cargo. For example, a temporary boiler or generator
- For container ships, particulars of the ownership and value of the containers on board and evidence of where insured
- Final freight account, showing the gross freight earned for the voyage and giving particulars of any freight pre-paid or advanced at the time of the event giving rise to GA.

If at the time of the event giving rise to GA, any freight that was at risk of parties other than cargo:

- Agents' or general accounts, with details of costs for any ports of call subsequent to the GA act, including ports of call
- If not included in the above, the cost of discharging the cargo at destination(s)
- Details of the bunkers consumed and remaining on board at the time of the event giving rise to GA
- Details of the bunkers consumed and remaining on board at the time of the final discharge of cargo
- Details of the cost of bunkers taken on after the GA act and before the final discharge of the cargo.

Policies of insurance on:

- Disbursements and/or increased value of the ship
- Excess liabilities
- Freight (including any special voyage freight policies).

Commercial insurance claims

The collection of evidence primarily concerns the interest of the owner, manager or charterer of a vessel. P&I Defence Clubs provide insurance for these claims but this is restricted to legal and other costs reasonably associated with the defence and pursuance of a commercial claim.

The insurance, subject to various provisos, covers new building warranties, port delays, vessel performance, port and berth safety, freight hire, demurrage and bunker disputes. Pages 72-78 of the handbook cover these disputes.

War risk claims

These insurance policies cover the same types of losses that would be insured by P&I, H&M and loss of hire (LOH) underwriters as a consequence of a war risk that includes war, capture, seizure (except piracy) and persons acting maliciously or with a political motive. However, following a recent case this does not generally include seizure or confiscation of the vessel due to smuggling or customs infringement, even if this happened without

Chapter 8
Evidence for insurance claims

the knowledge of the crew. Limited cover is provided by the P&I clubs as smuggling or customs infringement is generally not a 'war risk.' However, terms and conditions can be amended to remove this loop hole.

These claims, although infrequent, have arisen from:
- Ships trapped in the Suez Canal
- Gulf War damage
- Terrorism damage.

Collection of evidence for these claims would be broadly the same as for other claims, depending on the loss or liability. In the event of war or a terrorist act, this may present some unique situations that are beyond the scope of this chapter.

image: Danny Cornelissen (www.portpictures.nl)

THE NAUTICAL INSTITUTE

Chapter 9

A practical view from P&I

By Louise Hall

Protection & Indemnity (P&I) Clubs mutually insure their shipowning and chartering members' liabilities arising out of the operation of their ships. Claims can include, but are not limited to, liabilities relating to personnel (crew and passengers); collision; damage to, and loss of, property; cargo-related issues; and pollution.

While the ship's crew work in close conjunction with the insured member's shore management towards a common goal of conducting shipboard operations safely and in compliance with the relevant and prevailing statutory regulations, unfortunately incidents do occur. These incidents expose members and the Club to claims, some of which may be substantial in terms of the costs involved and the liabilities incurred.

The ability of the Club to defend the position of the member and their claims depends heavily on the documentary evidence and records that are made available to the Club and the team working on the incident. This may include lawyers, correspondents and surveyors, for whom the ship's crew plays a vital role at the time of an incident. Appropriate action with regards to collecting, preserving and presenting necessary evidence may help protect the member and assist in mitigating potential claims.

Case study 1

Strainer box gaskets on board a deepsea vessel burst while a bunker tanker delivered gas oil, causing a spill on the deck and into the sea. The deepsea vessel was fined by the port authority, but alleged that the bunker barge had exceeded the agreed loading rate and sought indemnity for this and the cost of cleaning the vessel.

Fortunately, the crew on board the bunker barge had kept efficient records of the transfer operation and were able to show quite conclusively that the agreed loading rate had never been exceeded. On investigation, it was concluded that the incident was likely to have occurred due to the increase in pressure resulting from the valves being closed on the deepsea vessel, causing the packing to burst.

Observations
This case highlights the need to keep proper records during oil transfer operations. Had the barge's crew not been diligent in recording events, taking regular tank soundings and noting pump speeds, it would have been much more difficult to avoid liability.

Financial cost
Nil

Chapter 9
Guidelines for Collecting Maritime Evidence

If effective evidence is not collated at the time of the incident this may be detrimental to the member's defence. Some claims may take years to be lodged and it is unfortunately often the case that the fact that evidence is insufficient is not discovered until the official notification is made.

Case study 2

A passenger fell into an open hatch on board a 40gt passenger excursion vessel. At the time of the incident, the vessel was in port and the majority of passengers had been asked to wait on the quay while repairs to the engine were being carried out. A few passengers, who were the official organisers for that trip, were already on board from the previous port and had been warned by the vessel's crew of the hatch being left open to the engine room in order to progress the repairs.

One passenger from this group was on the quay side and was called on board by his fellow organisers. Despite prior warning from the Master, the passenger rushed on board, brushed past the deckhand guarding the open hatch, and subsequently fell into the space below. After the incident, the ship's crew offered medical assistance. The passenger refused this and was observed to enjoy the rest of the cruise.

During subsequent court proceedings, the passenger alleged that the owner and his crew on board were negligent in their duty and had exposed him to a risk of injury.

In their initial statements made at the time of the incident, the owner, Master and deckhand had stated that the passenger had forcefully boarded the vessel, having jumped a small 1ft gap between the dock and the vessel as the gangway had not been arranged. However, during the court proceedings, the Master offered a different recollection stating that the passenger had boarded the vessel from the gangway, having been granted permission to board.

Although the Master's statement tallied with those of the owner and the deckhand in all other aspects, the difference in the statements about granting of permission to board and the positioning of the gangway substantially weakened the case. Finally, a settlement was reached.

Observations
Had evidence been preserved at the time of the incident, the claim may not have substantiated to the level it eventually did. Evidence that could have been beneficial includes:

- Statements from the owner, Master and deckhand duly recorded, signed and even notarised, in case a claim was raised after a considerable amount of time. Statements recorded at the time of the incident hold greater value than statements made later based on recollections and memory
- Statements from the owner, Master and deckhand should have been recorded in the deck log book or official log book, together with a detailed description of the incident

- Statements from the passenger affected should have been recorded and signed, including his refusal to accept the offer of medical assistance
- Statements of other passengers waiting on the quay should have been taken to establish whether the claimant had jumped on board and made his way in forcefully
- Photographs of the notices and warning signs in place to warn of the open hatch
- Log book records of all warning announcements made and any pre-work checks undertaken, including lockout or tag out procedures, plus copies of relevant checklists that had been complied with.

There were also allegations that the claimant was under the influence of alcohol at the time of the incident. This could not be established due to lack of evidence.

This case highlights the importance of preserving evidence for every incident, especially where passengers are involved, irrespective of how minor the incident appears at the time.

Financial cost
$393,137

Contemporaneous evidence is particularly important in incidents involving personal injury, as claims may be presented many months or years after the alleged incident took place.

The one potential exception to the general rule that contemporaneous evidence collected or recorded by the crew is beneficial is in significant casualty situations. In such cases, the primary recommendation to the Master and crew is to preserve existing evidence and immediately contact the nearest local correspondent who will be able to appoint legal representation that will ensure that any statements or incident reports are afforded 'litigation privilege' (if possible in the relevant jurisdiction). By doing so, any evidence that is potentially prejudicial to the insured member will not automatically be disclosable to third parties. See Chapter 6 for further details.

In addition to assisting in the defence of a claim by trying to establish the events leading up to the incident, of equal if not more importance, is establishing the perceived cause for the occurrence. P&I Clubs and industry standards such as the ISM Code support that learning from past incidents can only improve maritime safety.

One of the proactive measures highlighted by the ISM Code is the analysing of accidents, incidents and near-misses to create a safety environment that encourages learning from the dissemination of this open information.

Vessels below 500gt are not required to conform to SOLAS or the ISM Code. Nor are vessels trading domestically where their flag state has not adopted these requirements. However, even if it is not a regulatory requirement, it is advisable to implement a procedural system to encourage an adequate level of onboard safety and quality. It acts as a measure of good practice and demonstrates that a company aims to raise the bar on safety, as well as assisting in preventing damage and harm to crew, vessels and the environment.

Chapter 9
Guidelines for Collecting Maritime Evidence

General evidence (see the checklist pp 9-12 in *The Mariner's Guide to Collecting Evidence – Handbook*, hereafter referred to as the handbook) must be collated for every incident.

What follows is taken from the author's own experience of how one particular P&I Club reviews claims that are notified to it. To assist with the claims handling and causation process, the Club will focus on:

- Reports, including accident investigation reports filed with flag state, port state or other authority
- Statutory certificates and certificates issued by the classification society, including any current memorandums of class
- Ship's drawings or plans if relevant to the incident
- Verification of manning and rest hours in the form of a Safe Manning Certificate, crew certificates of competency and flag state endorsements, records of hours of rest/work and the current crew list
- VDR, S-VDR to prevent overwriting of data
- Photographs and CCTV footage
- Company procedures and checklists in place appropriate to the incident
- Log book entries
- Documentation, records and evidence, including defective equipment, RADAR, ECDIS and other records relating to the incident
- Photos or videos of any damage, injury or condition relating to the incident, preferably with date and time displayed
- Statements of facts from crew members and any other witnesses. These should be signed by the person concerned and counter-signed by the personnel taking the statement.

In all cases, a primary claim type is assigned that governs what further details need to be gathered:

- Cargo damage or shortage – were the prevailing weather conditions a factor? What was the causal reason for poor handling or shortage?
- General (including fines and stowaways) – assessment of the procedures in place under the ISPS Code or Ship Oil Pollution Emergency Plan (SOPEP)
- Injury or illness – type of injury or illness; location of injury; was the illness pre-existing? Was a pre-employment medical undertaken before joining the current vessel?
- Navigation – assessment of whether machinery failure, handling error, failure to adhere to the Colregs or weather conditions were factors
- Pollution – assessment of why the pollution occurred: failure of procedures, human error or as the result of a collision?
- Wreck removal – details of the removal and steps taken to mitigate against further risks, such as pollution.

In addition, every claims incident notified to the Club is assessed for primary cause and therefore further causational specific criteria need to be collated by an appointed correspondent or attending surveyor, depending on the type of incident. With each assessment is the further requirement for a primary cause to be assigned from one of nine set criteria:

- Third party negligence
  A cause deemed to be entirely outside the control of the member and ship's crew

- Inadequate procedures or guidance
  When the ship's operator fails to put in place procedures or operational guidelines for crew members to follow. For this to be a primary cause it has to be established that these were not in place, otherwise the onus would be on the crew member

- Act of God
  An event that is beyond human control; a naturally occurring phenomenon such as a rogue wave or a bolt of lightning hitting a ship

- Insufficient training
  This could be onboard training in the use of a specific type of machinery or a general lack of training (eg poor seamanship, or crane driving skills)

- The human factor
  This is a primary cause when the incident was brought about by the failure of a crew member to take the reasonable care that a prudent person would be expected to exercise when undertaking a specific task. It may involve the crew member failing to follow procedures, instructions or making an error in judgement (such as taking the correct course of action but failing to appraise the situation fully). A situation may arise whereby data given to the crew member is not acted upon correctly. Routine can also be a cause, whereby a task is repeatedly undertaken and for no apparent reason a deviation from the norm occurs

- Personal illness
  Akin to an Act of God (see definition above). When personal illness gives rise to a claim rather than an accident or other physical cause

- Inadequate ship maintenance
  Poor or no maintenance regime in place

- Inadequate shore-based management
  Claims caused by the direct or implied action or non-action of shore-based management. It can take into account failure to provide assistance to the vessel after it had been requested, such as a failure to provide requested spare parts or make up a deficiency in crew numbers

- Other
  This includes claims from several areas, including incident notifications originating from another Club within the International Group of P&I Clubs known as a Pool claim, a claim regarding a legal or contractual dispute or a steel survey or other pre-shipment survey.

The collation of this pertinent causation analysis enables the Club to detect regional, vessel-specific, sector or member-specific trends concerning their operation. This complements the Club's own analysis and can be channelled into suitable initiatives

Chapter 9
Guidelines for Collecting Maritime Evidence

such as educational booklets and bespoke training. In addition, the dissemination of case studies to the rest of the Club's members and the industry as a whole assists with the 'lessons learnt' process akin to the requirements of an efficient management system and of the ISM Code.

The overall aim of these initiatives is to assist vessel operators to mitigate against further claims incidents and thereby prevent similar incidents from happening in the future. From a P&I point of view, such initiatives also help to lower the number of claims incidents a member and the Club as a whole incur. This benefits and improves the loss ratios for the member and Club, which in turn has a positive effect on associated premiums.

Chapter 10

Evidence and arbitration

By Captain Julian Brown

Arbitration is a private process in English law, separate from the courts. Outcomes are rarely made public, which is one of the reasons why many people in shipping favour it as a way of resolving differences. However, that very confidentiality hints at why guidance on the use of evidence in arbitration is necessary.

Arbitration proceedings are (with limited exceptions) not published and neither are arbitration awards. For those outside the tight-knit world of maritime arbitration disputes the process can be a bit of a mystery. Therefore, this chapter will look at the most common rules and terms used for maritime arbitrations; how evidence is used; provide some background and an introduction to what is expected of factual and expert witnesses; and discuss examples of how following best practices on board can influence how a tribunal views the evidence before it.

Arbitration rules and procedures

The ways in which a shipping dispute can find its way into arbitration are numerous. Typically, there is a dispute between an owner and charterer or receiver for demurrage, freight or damage to cargo. Shipbuilding, offshore and salvage disputes are also commonly settled by arbitration. The charterparty will usually specify the forum of the dispute, the terms or rules agreed upon and perhaps the applicable law.

Many maritime arbitration disputes are dealt with under the London Maritime Arbitrators Association (LMAA) Terms. The LMAA terms provide for arbitration under the English 1996 Arbitration Act.

An important characteristic of arbitration is the flexibility it provides the parties. The arbitrator will determine whether to apply any rules of evidence, whether to receive hearsay evidence, and will see documentary evidence such as log books, drawings, plans, charts.

In practice arbitrators tend to err on the side of caution and to admit most statements or documents: excluding evidence increases the risk that an aggrieved party will appeal to the courts. The relevance or weight that an arbitrator gives to any particular document is another matter, however.

Institutional rules generally give tribunals and parties wide discretion to determine the extent to which evidence is presented and treated. There are significant differences in the way common law and civil jurisdictions treat document production and disclosure.

Chapter 10
Guidelines for Collecting Maritime Evidence

This is recognised in the International Bar Association (IBA) Rules: these provide a formal mechanism for the presentation of evidence which the parties can agree to adhere to on a voluntary basis, for the purpose of arbitration.

Key elements of the IBA rules are:

Article 4. Witness statements to contain:
- The full name and address of witnesses
- Relationship (if any) to any of the parties
- Description of their background, qualifications, training and experience, if relevant to the dispute or to the contents of the statement
- Full and detailed description of the facts, and the source of the witness's information
- Additional documents witnesses rely on
- Language the witness statement was originally prepared in and the language in which witnesses propose to give evidence
- Affirmation of the truth of witness statements
- Signatures of witnesses and date and place of signature.

If a witness fails to appear without a valid reason at an Evidentiary Hearing, the Arbitral Tribunal can disregard any statement they have made unless it decides otherwise. This is important and challenging in shipping, as ship's staff who are material witnesses may be unavailable working at sea, uncontactable when on leave or no longer in the employ of the owner.

Article 5. Requirements of expert's report:
- Full names and addresses of experts
- Relationship (if any) to any of the parties, their legal advisors and the tribunal
- Their background, qualifications, training and experience
- Instructions about opinions and conclusions they are to provide
- Statement of their independence from the parties, their legal advisors and the tribunal
- Statement of the facts on which they are basing their expert opinions and conclusions
- Description of the methods, evidence and information used in arriving at their conclusions
- Submission of new documents not already provided
- Statement of original language if the expert report has been translated
- Language in which the party-appointed expert anticipates testifying
- Affirmation of their genuine belief in the opinions expressed in the expert report
- Signatures of the party-appointed expert witnesses and date and place of signature
- If more than one expert has signed the report, the specific parts of the expert report must be attributed to each author.

The tribunal can direct party-appointed experts to meet and reduce the areas of disagreement. Witnesses (factual and expert) need to appear in person unless the tribunal allows the use of videoconference or similar technology.

Chapter 10
Evidence and arbitration

Article 9. Admissibility and assessment of evidence
Evidence might be excluded if:
- It lacks relevance to the case
- It is found to be subject to legal privilege
- Producing the evidence would be an unreasonable burden
- Documents are lost or destroyed
- The tribunal determines there is compelling commercial or technical confidentiality.

Legal representatives from all parties should highlight to both factual and expert witnesses the importance of the confidentiality of the proceedings.

The LMAA terms

The LMAA is an association of practising arbitrators. It publishes a set of terms for the conduct of arbitrations. These terms are those on which LMAA members usually accept arbitration appointments in maritime cases.

A new Fourth Schedule (previously published as a separate checklist) has been added to the LMAA terms (2017). This contains important guidance that should make the process more efficient and reduce costs. If a party fails to comply with the guidance, they could find that the tribunal makes them pay more for costs if that failure has delayed the progress of the arbitration or cost the other party more money.

Witnesses

Witnesses of fact should only provide evidence of what they heard, what they saw and what they did. They should not provide their opinions as to fault or cause – this is for the tribunal to determine. An expert witness is a person who provides an independent expert opinion and whose duty is to assist the tribunal. Expert witnesses may give both opinion evidence within their expertise and evidence of facts. Expert witnesses should have appropriate qualifications and experience in their professional fields and undertake recognised expert witness training.

Mediation

A mediator, who need not be academically or professionally qualified, is a neutral party appointed to help achieve resolution between parties. Mediators are not empowered to settle disputes and there is generally no binding outcome in mediation unless there is an agreement, typically reduced to writing. Mediation is distinct from arbitration, although it may run in parallel with arbitration proceedings. Mediation may or may not be held under the auspices of an institution, is more informal and less structured than arbitration. It is not necessary for evidence to be tendered during the mediation sessions and there is no examination or cross-examination.

Chapter 10
Guidelines for Collecting Maritime Evidence

Nevertheless, a good deal of factual and expert evidence is likely to have been prepared by both parties to demonstrate the 'soundness' of their respective cases. In the event that the matter does not settle at mediation, the case will typically revert to arbitration or court.

Shipboard evidence

One of the most difficult challenges that a tribunal has to grapple with is the quality of shipboard produced evidence. Expert reports are meant to assist the tribunal but they often fail to do so. It is not uncommon to find experts interpreting evidence in different ways so issues become less clear. That can be particularly so in very technical disputes or where a tribunal does not have the right technical expertise represented. Following good procedures and industry practices can make all the difference as to how tribunals view shipboard evidence.

These typical examples of two similar sized chemical tankers (Ship 1 and Ship 2) are taken from the chemical tanker trade where cargo quality disputes are not uncommon – although the same broad principles apply to other ship types. Receivers will typically claim that owners are to blame for causing the cargo to go off specification, pointing to the clean bill of lading as *prima facie* evidence that the carrier is at fault. Owners often claim that the cargo was off specification on loading, that cargo tanks were properly prepared and the fault lies with the shipper.

The case studies below show the difficulties that conflicting evidence poses for tribunals. Many similar disputes could be resolved far more speedily and cost effectively if good cargo sampling practices were adopted.

Case study 1

Ship 1 loaded a cargo of vegetable oil. On completion of loading the cargo, surveyors took samples from each cargo tank loaded, drew composite samples and placed these together with shore tank samples on board Ship 1 for delivery to the receivers. On arrival at its discharge port, cargo surveyors analysed the cargo and found it to be off specification, and it was rejected. The vessel was instructed to leave its berth. The dispute went to arbitration. Owner's representatives were ultimately unsuccessful in defending the claim for off specification, in part due to the limited evidence available.

Case study 2

Ship 2 also loaded a cargo of vegetable oil. On completion of loading the cargo, surveyors took samples from each cargo tank loaded, drew composite samples and placed these together with shore tank samples on board Ship 2 for delivery to the receivers. However, Ship 2 had good company procedures and a focused crew, aware of the commercial risks inherent in the trade. Ship 2 staff took their own samples – from all cargo manifolds in use before loading; from those same manifolds at the start of, and periodically during, loading; and first foot samples. An alert chief mate

> obtained a sample from a shore cargo hose before connection. Samples were kept in appropriately labelled self-sealing bottles, marked with the port, date, time, location and grade etc. Suitable entries were made in a ship's sample log. They took the added precaution of endorsing the cargo surveyor's sample report (cargo surveyors are generally unwilling to acknowledge receipt of an equivalent document issued by the ship) with the relevant ship-taken sample information. On arrival at the discharge port, cargo surveyors analysed the cargo and found it to be off specification, and it was rejected. The vessel was instructed to leave its berth. The dispute went to arbitration. The comprehensive samples and detailed contemporaneous evidence provided by Ship 2 enabled the owners to successfully defend the claim for off specification and demonstrate that it was caused by previous cargo residues in the shore lines and hoses.

In these two examples only the Ship 2 claim was successfully defended. The Ship 1 claim dragged on for more than two years and was ultimately costly for the owners. The different outcomes were almost entirely due to the evidence collected on board.

Involved on that (costly) journey were owners, operators, receivers, charterers, shippers, P&I insurers, cargo underwriters, lawyers, chemists, experts and surveyors. All were endeavouring to answer the ultimate and deceptively simple question – what was the cause? And of course its result – who pays?

Similar procedures, modified as appropriate, apply to fuel oil and chemical cargoes. Some cargoes may be more complicated or hazardous to sample than others and a suitable risk assessment, combined with appropriate procedures and PPE, need to be identified before each such operation. Other examples abound of where failure to follow good industry practices have resulted in unnecessary and frequently costly disputes – whether sampling solid bulk cargoes, precautions when line pigging or blowing, or taking shore nitrogen.

Conversely, demonstration of compliance with industry and best practices can frequently provide a successful defence.

There is a wealth of best practice literature available. In the cases of chemical and gas tankers, the Vessel Inspection Questionnaires (VIQs) published by the Chemical Distribution Institute (CDI); for tankers of all types the Oil Companies International Marine Forum (OCIMF) SIRE VIQs. In terms of management best practices, OCIMF's Tanker Management Self Assessment (TMSA3) is generally regarded as the gold standard. While primarily designed for tanker operators, its recommendations can be applied, with modifications, to any fleet type.

The bulk carrier industry has the Rightship system to promote ship quality. The Nautical Institute, SIGGTO, OCIMF and ICS among others all have excellent industry publications whose contents should guide every owner's management systems. And almost all P&I Clubs have loss prevention programmes and guides dealing with everything from bunker issues to crew claims to navigation practices.

Chapter 10
Guidelines for Collecting Maritime Evidence

Practical guidance for the collection of maritime evidence

- Record immediately everything that might relate to a problem present or future – even when no problem is known at the time. In a recent case an officer made a video and took photographs of seas at different stages of a voyage, which were invaluable in an arbitration. Damage was said to be due to weather, although the seas were moderate; his evidence made it clear that the damage was due to bad stowage – which he had also photographed during loading by the charterers' stevedores!

- When taking photographs make sure the time and date function is activated. If possible, make a note of who took the photographs and where from.

- It is often best to let the crew record their own versions of events and collect them. However, if it is necessary to interview a crew member, try to do this word for word – ideally in a recording – noting the time and date when the evidence is recorded.

- Make sure the names and full contact details of witnesses are recorded and kept for the future. Ensure that these records are kept safe and can be found.

- All log books – including deck, engine and cargo – should be kept in ink and accurately reflect the events they describe. It is not acceptable to exaggerate or even deliberately falsify entries about weather encountered on voyage.

- Tribunals may consider evidence differently from other forms of litigation such as courts and adjudication, but they have a similar philosophy about witnesses. Both expert and factual witnesses are expected to give honest and truthful evidence that will assist the tribunal in resolving the dispute.

Appendix

The no blame approach to state safety investigations

By Captain Paul Drouin

So important is the blame-free approach to a state safety investigation that most investigative agencies have gone to the trouble of indicating this principle in their published reports. Examples of some countries or authorities that undertake safety investigations within the intent and spirit of the IMO Casualty Investigation Code are listed here, together with the wording of the proviso found in their respective reports.

Country or authority	Proviso
Transportation Safety Board of Canada	The Transportation Safety Board of Canada (TSB) investigated this occurrence for the purpose of advancing transportation safety. It is not the function of the Board to assign fault or determine civil or criminal liability.
Marine Accident Investigation Branch (UK)	The sole objective of the investigation of an accident under the Merchant Shipping (Accident Reporting and Investigation) Regulations 2012 shall be the prevention of future accidents through the ascertainment of its causes and circumstances. It shall not be the purpose of an investigation to determine liability nor, except so far as is necessary to achieve its objective, to apportion blame.
Australian Transport Safety Bureau	Readers are advised that the ATSB investigates for the sole purpose of enhancing safety. Consequently, reports are confined to matters of safety significance and may be misleading if used for any other purpose.
Bureau d'enquêtes sur les accidents de mer (France)	The analysis of this incident has not been carried out in order to determine or apportion criminal responsibility nor to assess individual or collective liability. Its sole purpose is to improve maritime safety and the prevention of maritime pollution by ships. The use of this report for other purposes could therefore lead to erroneous interpretations.

Appendix
Guidelines for Collecting Maritime Evidence

The Hong Kong Special Administration Region Marine Department Marine Accident Investigation Section	The purpose of this investigation ... is to determine the circumstances and the causes of the incident with the aim of improving the safety of life at sea and avoiding a similar incident in future. ... They are not intended to apportion blame or liability towards any particular organization or individual except so far as necessary to achieve the said purpose.
National Transportation Safety Board (USA)	The Independent Safety Board Act, as codified at 49 U.S.C. Section 1154(b), precludes the admission into evidence or use of Board reports related to an incident or accident in a civil action for damages resulting from a matter mentioned in the report.
Dutch Safety Board	The aim in the Netherlands is to limit the risk of accidents and incidents as much as possible. If accidents or near accidents nevertheless occur, a thorough investigation into the causes, irrespective of who are to blame, may help to prevent similar problems from occurring in the future. It is important to ensure that the investigation is carried out independently from the parties involved.
Danish Maritime Accident Investigation Board	The investigations are carried out separately from the criminal investigation, without having used legal evidence procedures and with no other basic aim than learning about accidents with the purpose of preventing future accidents. Consequently, any use of this report for other purposes may lead to erroneous or misleading interpretations.
Japan Transport Safety Board	The objective of the investigation conducted by the Japan Transport Safety Board… is to determine the causes of an accident and damage incidental to such an accident, thereby preventing future accidents and reducing damage. It is not the purpose of the investigation to apportion blame or liability.
BSU (Germany)	The sole objective of this investigation is to prevent future accidents and malfunctions. This investigation does not serve to ascertain fault, liability or claims. This report should not be used in court proceedings or proceedings of the Maritime Board.

Marine Safety Investigation Unit, Malta Transport Centre	This safety investigation report is not written, in terms of content and style, with litigation in mind and ... shall be inadmissible in any judicial proceedings whose purpose or one of whose purposes is to attribute or apportion liability or blame, unless, under prescribed conditions, a Court determines otherwise.

As can be seen, and with the possible exception of Hong Kong that has a slight dip towards blame, each authority has taken great pains to specify the no-blame approach. Even so, the Casualty Investigation Code is quite clear:

It is not the intent of the Code for a State or States conducting a marine safety investigation to refrain from fully reporting on the causal factors of a marine casualty or marine incident because blame or liability may be inferred from the findings.

Nonetheless, the Code also stresses that states involved in a marine safety investigation should ensure that marine safety records in their possession are not disclosed in criminal, civil, disciplinary or other administrative proceedings except in certain circumstances. These mostly concern the public interest in the administration of justice.

image: Danny Cornelissen (www.portpictures.nl)

THE NAUTICAL INSTITUTE

Index

A

Accident
- logs 8–9, 15, 17
- notifications 8, 15, 21, 24
- reporting forms 17–21
- reports 7–8
- scene investigation 6–7, 29–30, 66–67

Admiralty and Commercial Court, London 39

AIS (automatic identification system) 2, 28, 31, 35, 36, 40–41, 42

Arbitration
- dispute resolution 25, 71
- London Maritime Arbitrators Association (LMAA) Terms 71, 73
- mediation 73–74
- rules and procedures 71–73
- shipboard evidence 74–75
- tribunals 25, 26, 71–73, 74–75, 76

Atlantik Confidence 28

Audio recordings 16, 35, 42–43

Average adjustment 48–49, 58

B

Best practice literature 75

Bunkering 65

C

Cargo 51
- case study 52
- dry cargo 55–56
- general average 49–50, 52, 61–62
- liquid cargo 55
- P&I insurance claims 54–56

Cargo heating 52, 55

Case studies
- *Atlantik Confidence* 28
- bunkering 65
- cargo 52
- machinery failure 27, 49
- passenger injury 66–67
- tribunal evidence 74–75

Casualties, personnel
- accident/incident reporting forms 17–19
- crew personal injury, fatalities and illness 56
- prioritising 9, 47

Casualty Investigation Code 5, 6, 77, 79

CCTV systems 10, 14, 35, 36, 68

Charterers, P&I Clubs and 23, 26

Collisions
- evidence specific to 24, 28, 33, 51, 60
- insurance claims 57, 58

Colregs (International Regulations for Preventing Collisions at Sea) 28, 68

Commercial correspondents, P&I Clubs 24–25

Common law 39, 45, 71

Contact with fixed and floating objects (FFO) 24, 34–36
    P&I insurance claims 57

Contemporaneous evidence 31, 40, 41, 67, 75

Costa Concordia 55

Crew lists 14, 32, 68

Crew personal injury, fatalities and illness 56

D

Deviation to save life at sea 56

Disclosure of documents 27, 41–42, 45, 67

Dispute resolution 25, 47

Documents
    disclosure 27, 41–42, 45, 67
    general average 61–62
    insurance claims 48–9, 50–51, 58–63, 66–67, 68, 76
    photographing 31

Drug and alcohol testing 9

Dry cargo 55–56

Due diligence 44, 45–46

E

ECDIS (electronic chart display and information system) 2, 10, 32, 33, 41

ECS (electronic chart system) 10, 14

Electronically acquired evidence 2–3, 9–10, 14, 28, 31, 33–34, 40–41
    analysing 42–43

Engine room logs 3, 58–59

Equipment, involved in incidents 16

Evidence *see* Maritime Evidence

Expert witnesses 28, 71, 72, 73

Explosions 58

F

Fires 43–44, 51, 58

First aid treatment, records of 15

G

GA (general average) 49–52, 61–62

Groundings 34, 40, 41, 51, 58

H

Hebei Spirit 54

Herald of Free Enterprise 5, 55

H&M (hull and machinery) insurance
    accident/incident reporting forms 19–21
    collision liabilities 51
    insurance claims 57–62
    P&I Clubs, liaising with 51

I

Incidents
    causal factors 68–70, 75
    incident reporting forms 17–21
    navigation incidents 14, 16, 28, 30–37
    notifications 8, 15, 21, 24
    see also Accident

Insurance
    average adjustment 48–49, 58
    claims documentation 48–9, 50–51, 58–63, 66–67, 68, 76
    commercial insurance claims 62
    evidence for claims 53–63
    fraud 43–44
    H&M (hull and machinery) claims 57–62
    indemnities 47, 49, 65
    insurers, evidence required 48
    laws covering 53
    P&I insurance claims 53–57
    terms and conditions 53
    types of 47
    war risk claims 62–63
    warranties 44, 53, 60

Insurance Act 2015 53

International Bar Association (IBA) Rules 72

IMO (International Maritime Organization)
    Casualty Investigation Code 5, 6, 77, 79
    electronically acquired evidence 3
    mariner's liabilities 53
    VDR regulations 2

ISM (International Safety Management) Code
    accident analysis 67
    auditing 33, 44, 45
    compliance 29
    DOC (document of compliance) 49, 60
    electronic engine room logs 3
    vessels exempt from 67

L

Lawyers
    evidence and 39–46
    general average 50
    P&I Clubs and 24–25, 27
    see also Mariner/lawyers

Legal privilege 26, 27, 41–42, 67, 73

Liabilities
    assessing 24, 25
    charterers' 26
    mariner's, IMO and 53
    passenger incidents 54–55, 66–67
    P&I Club members' 23, 24, 51, 53–54, 65
    seaworthiness affecting 44, 50
    shipowners' 23, 44, 45–46, 53

Lighting, relevance to incident 16

Liquid cargo 55

litigation 25, 26, 27, 41, 43, 45, 47

Litigation privilege 67

London Maritime Arbitrators Association (LMAA) Terms 71, 73

M

Machinery failure
    case studies 27, 49
    evidence and documentation 49
    insurance claims 58
    lawyers and 43–44
    surveyors 30

Manoeuvring 35–36

MAIB (Marine Accident Investigation Branch) 5–8, 77

Marine claims brokers
    collecting evidence, perspective on 47–52

Marine Insurance Act 1906 53

Mariner/lawyers 39–40
    disclosure of documents 27, 41–42, 45, 67
    electronic evidence 40–1, 42–43
    machinery damage and fire 43–44

Index

Maritime evidence
　arbitration and 71–76
　collisions, specific to 24, 28, 33, 51, 60
　contact with fixed and floating objects (FFO) 34–36, 57
　contemporaneous 31, 40, 41, 67, 75
　definitions 1
　electronically acquired 2–3, 9–10, 14, 28, 31, 33–34, 40–41, 42–43
　groundings 34, 40, 41, 51, 58
　for insurance claims 53–63
　insurers, collecting 48
　lawyer's point of view 39–46
　machinery failure 49
　manoeuvring 35–36
　marine claims broker's perspective on collecting 47–52
　Master's responsibilities for collecting 8–11, 13–21, 25–26, 67
　navigation incidents 14, 16, 28, 30–37
　non-perishable 11
　perishable 9
　photographic 10–11, 14, 17, 25, 31, 76
　physical 11
　P&I Clubs' approach to collecting 23–28
　practical guidelines for collecting 76
　preserving on behalf of state safety investigators 5–11
　securing 14, 21, 25–26
　shipboard evidence 74–75
　surveyor's perspective on collecting 29–37
　vessel resorting to a port of refuge 50, 61
　wash damage 36–37
　see also Documents; Witnesses

Master
　responsibilities for collecting evidence 8–11, 13–21, 25–26, 67

Media, dealing with 21

Mediation 73–74

Mooring 35–36

N

Navigation incidents 14, 16, 28, 30–37

Navtex 10

No blame marine safety investigations 5, 77–79

Non-perishable evidence 11

P

Passengers, insurance claims 54–55, 66–67

Perishable evidence collection 9

Permits to work 17

Photographs
　accident scene 10–11, 14, 17, 25, 31, 76
　of documents 31

Physical evidence 11

P&I Clubs
　accident/incident reporting forms 17–19
　collecting evidence, a practical view on 65–70
　collecting evidence, approach to 23–8
　evidence collection 23–24, 67–70
　H&M insurers, liaising with 51
　Insurance Act 2015 53
　insurance claims 53–57
　lawyers and commercial correspondents 24–25, 27
　surveyors and 26, 29, 54
　warranties 44, 53

Index

Pollution 54

Port of refuge, vessel resorting to 50, 61

PSC (Port State Control), advising of incidents 15

Proof, onus of 47

Proximate cause 8, 53

Q

Qualifications, records to be kept 17

R

Risk assessments
    records of 17

S

Safety investigations, no blame approach 5, 77–79

SMC (safety management certificate) 49, 60

SMS (safety management system)
    accident reporting forms 17–21
    MAIB investigations and 6–7
    reviews of 44

Salvage cases 61

Seaworthiness 44, 50

Securing evidence 14, 21, 25–26

Ship management
    advising managers of incidents 15, 21
    P&I Clubs and 23

Shipowners
    advising of incidents 15, 21
    liabilities 23, 44, 45–46, 53
    onus of proof 47
    P&I Clubs and 23

Social media 21

SOLAS
    no blame marine safety investigations 5
    vessels not covered by 67

Stowaways 56

Surveyors
    accident investigations 29–30
    collecting evidence, perspective on 29–37
    general average 50, 61
    navigation incidents 28, 30–37
    P&I Clubs and 26, 29, 54

T

Toolbox meetings 17

Training, records to be kept 17

Tribunals, arbitration 25, 26, 71–73, 74–75, 76

V

VDR (voyage data recorder)
    contact with fixed and floating objects 35, 57
    groundings 31
    IMO regulations 2
    preserving data from 9, 14, 28, 31, 42–43, 68

VHF records 14, 15, 28, 33, 42

VSMCs (very serious marine casualties) 5

VTS (Vessel Traffic Services) 28, 57, 58

W

War risk claims 62–63

Warning signs 17

Warranties 44, 53, 60

Wash damage 36–37

Weather, record of, at time of incident 16

Witnesses
    expert 28, 71, 72, 73
    identifying 14, 30
    interviews and statements 7, 9, 15–16, 26, 66–67, 72, 76
    reliability of witness evidence 26–28, 40, 41–42
    tribunals and 72–73

Contributors

Chris Adams BSc (Hons) FNI MRIN

Chris is a Partner and Director of the companies responsible for the management of the Steamship Mutual P&I Club, and currently Head of European Syndicate with responsibility for the business of all the Club's members domiciled in Europe, and Head of Loss Prevention with responsibility for all of the Club's loss prevention materials and initiatives. He is a Younger Brother of Trinity House and previously served at sea as a navigating officer with Ellerman City Liners on general cargo, container and ro-ro ships.

Captain Sanjay Bhasin LLM MNI

Sanjay is a partner at Solis Marine Consultants in London. He has sailed as Master on bulk, reefer and forest products carriers. After coming ashore, he worked in South Africa with P & I correspondents and a leading cargo insurer, investigating accidents and implementing loss prevention programmes. He moved to London in 2007 and was a director with a leading marine consultancy before joining Solis. He has wide experience of collecting evidence and investigating marine incidents and casualties and has given evidence as an expert witness.

Captain Julian Brown FNI FCIArb FCMS LLM

Julian has been arbitrating since 2011, mainly under LMAA terms, and has appeared as an expert witness since 1992. He has extensive experience of Singapore OPL operations, blending and contamination cases and related disputes. After a seagoing career to Master on a wide variety of vessels, he became an auditing OCIMF and senior CDI inspector for ships and terminals and a panel surveyor for the International Group of P&I Clubs. Julian is CEO of JCP Marine, which he founded in 1992.

Contributors
Guidelines for Collecting Maritime Evidence

Captain Paul Drouin FNI

Paul is currently the Principal of SafeShip.ca, a marine consulting firm specialising in marine investigations, safety culture and pilotage issues, and the editor of The Nautical Institute's Mariners' Alerting and Reporting Scheme (MARS). He spent 20 years as officer and Master with the Canadian Coast Guard and over a decade as a marine accident investigator with the Transportation Safety Board of Canada.

Ivor Goveas LLM MNI

Ivor is currently an Executive Director at Willis Towers Watson, providing claims advocacy to shipowners, charterers, ports and terminals. His sea-going career spanned 23 years, mainly on gas and petroleum tankers, rising to the rank of Master. After coming ashore, he worked in marine insurance before moving into marine claims broking where he has managed and mentored a claims team providing a full claims service and risk management solutions. Ivor is currently a member of The Nautical Institute's Professional Development Committee.

Louise Hall MSc Marine Surveying

Louise is currently Director, Loss Prevention at Shipowners' P&I Club, where she has global responsibility for developing and implementing the Club's strategy for loss prevention. In addition to serving at sea with a global container shipping line, she has experience of working as a cargo planner for both shipping lines and a cargo terminal, operations superintendent and as a ship manager.

Jack Hatcher MNI

Jack is a mariner and qualified English solicitor at Hill Dickinson LLP, specialising in the investigation and handling of marine casualties worldwide and related contractual and third party disputes. He has acted on a variety of wet and dry shipping matters for P&I clubs, hull insurers, salvors, owners, managers and charterers, in the English courts and in arbitration. Jack served at sea as a deck officer on passenger ships and has worked for an International Group P&I Club handling P&I and FD&D claims.

Contributors

David Keyes

David is Head of Marine Shipowners Claims at the global insurance broker, Willis Towers Watson, specialising in claims relating to hull and machinery and ports and terminals for clients worldwide. He has over 30 years' experience as a marine claims broker and client advocate in the London Insurance Market. He started his career with Sedgwick Ltd and then moved to the oil services company Halliburton, where he held the position of Risk & Insurance Manager for Europe, Far East & Africa.

Michael Mallin BSc HCMM MNI

Michael is a partner at the Hong Kong office of English solicitors Hill Dickinson LLP, having set up the office in 2013. As a Master Mariner with 12 years' seagoing experience, he specialises in problems arising from maritime casualties, including collisions, salvage, insurance and contractual disputes. During more than 30 years' practice as a shipping lawyer he has dealt with hundreds of casualties, including many of the largest (notably *Starsea, Costa Concordia, Chitra, Amadeo 1, Nissos Amorgos and APL Panama*), and has acted as an expert witness on many occasions.

Captain Andrew Moll BSc MNI

Andrew is Deputy Chief Inspector of Marine Accidents at the UK's Marine Accident Investigation Branch, having joined in 2005 as a Principal Inspector after 27 years in the Royal Navy. A surface and anti-air warfare specialist, his career was largely sea-going, spent in destroyers and aircraft carriers, with commands including the Type 42 destroyer HMS *York* and the Type 22 frigate HMS *Chatham*. He completed two appointments in the Ministry of Defence, in the Directorate of Naval Operations and as Secretary to the Chiefs of Staff Committee.

John Noble BSc FNI

John has been involved in many casualty cases as an expert witness and has significant experience of investigating marine casualties where evidence collection formed a core activity. His involvement with commercial shipping began in 1962 and he gained his Master's certificate in 1974. Between 1980 and 1999, John ran the consultancy Murray Fenton and Associates Ltd and continued as CEO when BMT acquired the business. John ran the Salvage Association for five years and now acts as a mentor to Constellation Marine Services in Dubai.

Contributors
Guidelines for Collecting Maritime Evidence

Captain Ian Odd AFNI

Ian was sailing on Panamax product tankers trading worldwide when he retired in 2013. He started his seagoing career in 1965 with a tramp company which then owned cargo ships and tankers and later acquired chemical tankers and shortsea vessels. After gaining his Master's certificate in 1976, he specialised in chemical tankers. His first command was in 1979. Employers were Stamford Tankers, AS Vulcanus and Expedo Ship Management. Ian is a volunteer member of the bridge team on SS *Shieldhall*, probably Europe's largest working sea-going steam ship.

Peter Young BSc MRINA MNI

Peter is currently Claims Director of V.Scope, the insurance broking arm of V.Ships. His sea-going career started on a cadet training ship, from which he qualified as a navigating officer with a BSc in Nautical Science. Employment as Office Manager with a tug and barge joint venture in Great Yarmouth servicing the developing offshore oil industry led to involvement in sorting out insurance and salvage claims. Peter subsequently worked for P&O's General Cargo and Bulk Shipping Divisions in senior claims management roles.

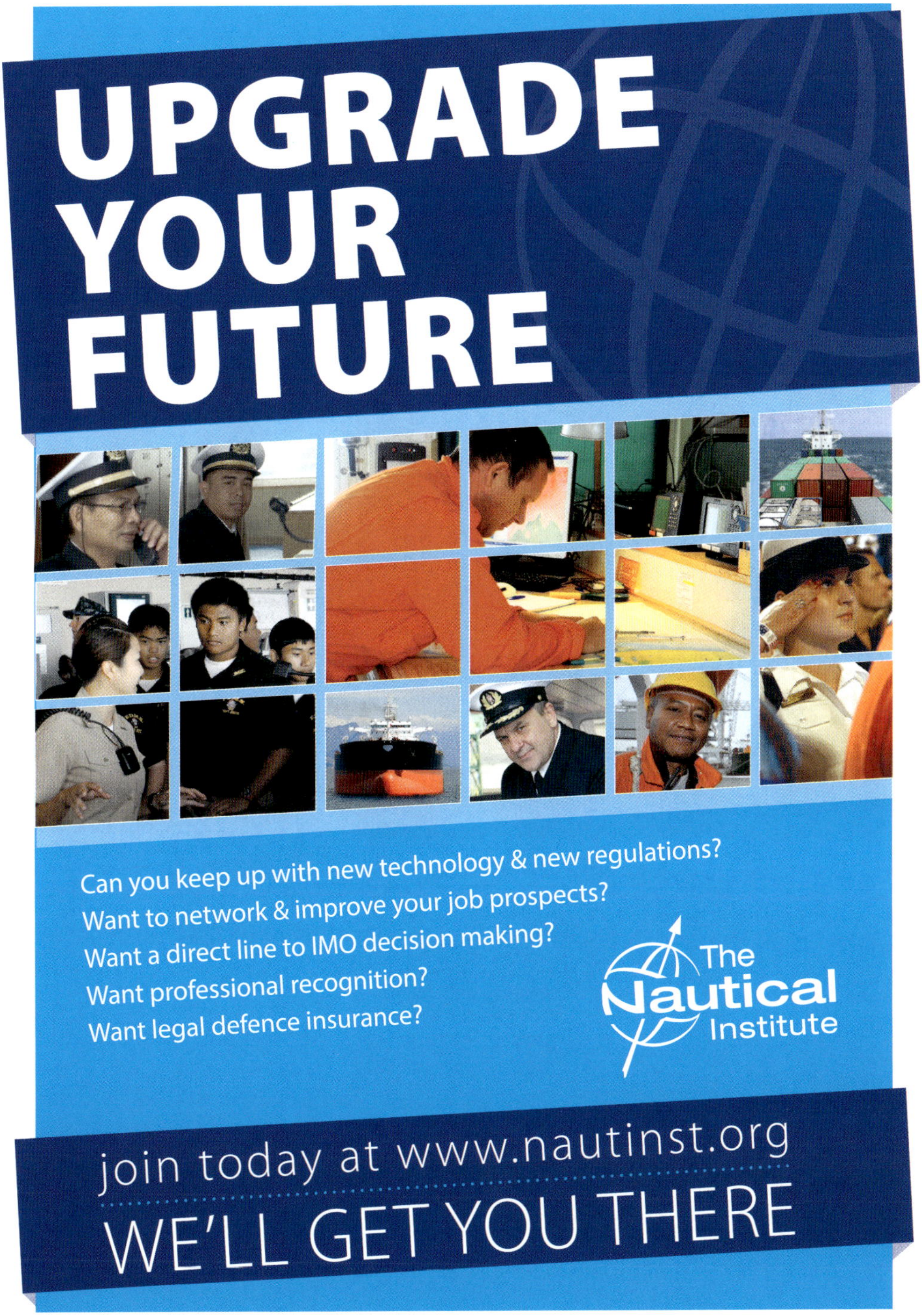
UPGRADE YOUR FUTURE
Can you keep up with new technology & new regulations?
Want to network & improve your job prospects?
Want a direct line to IMO decision making?
Want professional recognition?
Want legal defence insurance?
The Nautical Institute
join today at www.nautinst.org
WE'LL GET YOU THERE

Also available

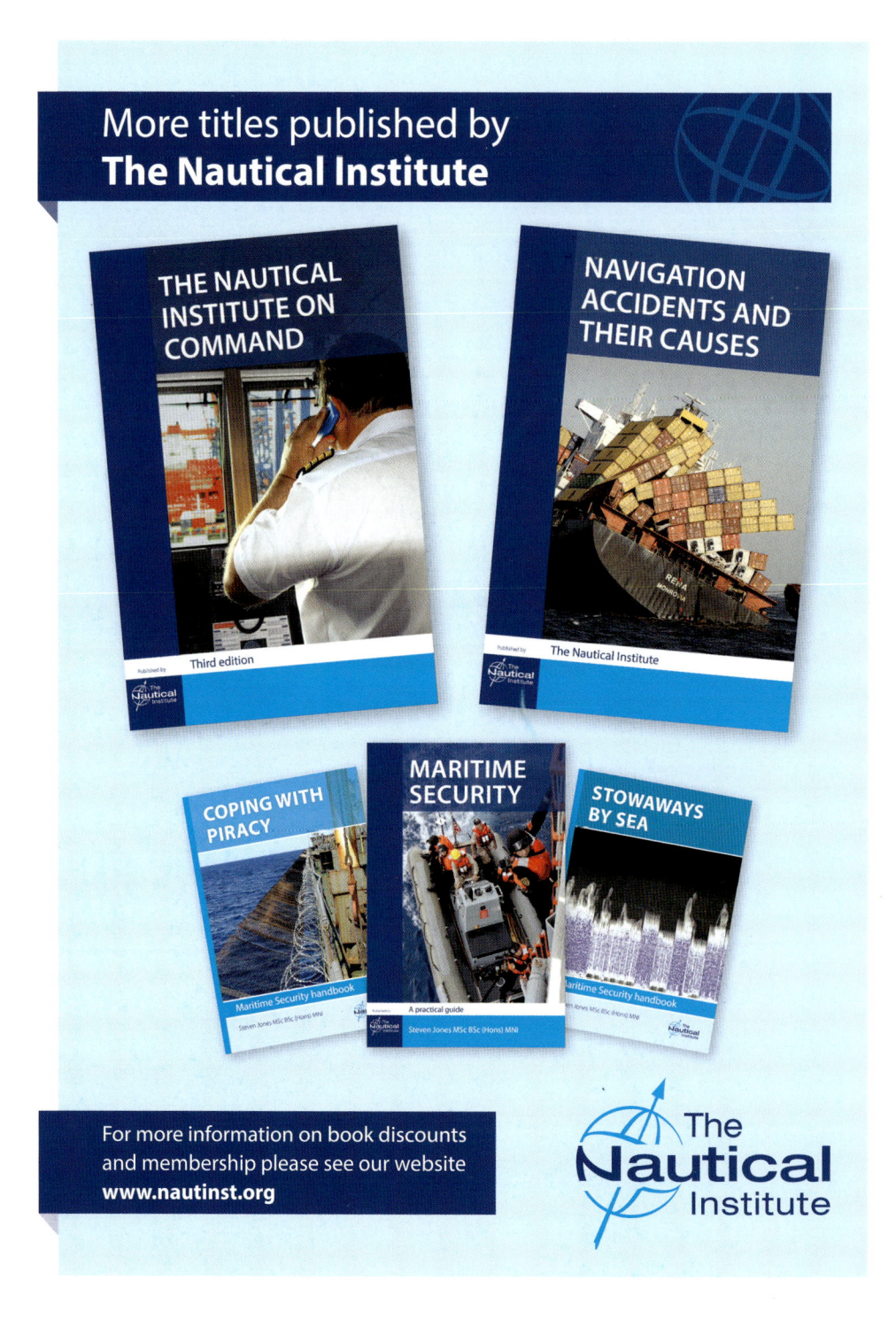

More titles published by The Nautical Institute

- THE NAUTICAL INSTITUTE ON COMMAND — Third edition
- NAVIGATION ACCIDENTS AND THEIR CAUSES
- COPING WITH PIRACY — Maritime Security handbook — Steven Jones MSc BSc (Hons) MNI
- MARITIME SECURITY — A practical guide — Steven Jones MSc BSc (Hons) MNI
- STOWAWAYS BY SEA — Maritime Security handbook

For more information on book discounts and membership please see our website
www.nautinst.org

The Nautical Institute